Modelling in Mechanical Engineering and Mechatronics

Nikolay Avgoustinov

Modelling in Mechanical Engineering and Mechatronics

Towards Autonomous Intelligent Software Models

 Springer

Nikolay Avgoustinov, Dr.-Ing.
Institute of Production Engineering/CAM
Saarland University
66041 Saarbrücken
Germany

British Library Cataloguing in Publication Data
Avgoustinov, Nikolay
    Modelling in mechanical engineering and mechatronics :
    towards autonomous intelligent software models
    1. Mechanical engineering - Computer simulation
    2. Automatic control - Computer simulation
    I. Title
    621'.0113
ISBN-13: 9781846289088

Library of Congress Control Number: 2007930212

ISBN 978-1-84628-908-8   e-ISBN 978-1-84628-909-5

Printed on acid-free paper

9 8 7 6 5 4 3 2 1

Springer Science+Business Media
springer.com

To Tania

Preface

Specificities of Modern Manufacturing

Nowadays, rapid and fundamental changes take place in government, business, technology and society. More than ever before, manufacturing is confronted with environment-related requirements. Globalization and eCommerce have led to the establishment of a demanding consumer market, where at least four general requirements for any single product concern manufacturers more than ever. These are high quality, low price, quick delivery, and last but not least – high customization, leading to an explosively increasing number of variants to be produced and to complexity of the products and their production. In these conditions of increasing competition, the trends to specialization and consolidation of small and medium-sized enterprises are a legitimate consequence.

One of the most challenging tasks related to the product and process modelling (as in many other areas) is the management of expert knowledge. This means that modelling is used in order to accelerate knowledge acquisition, to formalize knowledge representation and to enable knowledge transfer and reuse.
A key to achieving these goals is the proper application of modelling.

About the Book

History

Although I obtained a degree in mechanical engineering, my passion was actually automation and especially the use of computers for supporting it. The success of my first IT-project – my diploma work – was probably the cause for a shift in my professional career towards the interdisciplinary and new at that time field of application of computers for industrial automation. I have worked on many topics from very different areas – some of the more important are data exchange and conversion, integration of different applications or subsystems, virtual and mixed reality applications, architecture of software systems, 3D-visualization and simulation and even simulation and visualization for medical purposes. At some point in time, I noticed that all these topics have got much more in common than one usually supposes. In short, I have noticed that the efficiency of any solution in any particular field depends primarily on the quality of the models used at the beginning of each process chain. On the other hand, I have also noticed that some problem solving methodologies and tools specific to the separate topics influence the modelling too much. So much so that in extreme cases the experts begin to think in tool-related concepts or notions, and sometimes even forget that they are using models – with all the related consequences. Actually, something similar has happened to me. During the preparation of my PhD thesis about the exchange of product model data among a large number of CAx-systems I was really concentrated on these systems, on the respective standards for exchange and as it seemed to me – on the problems of exchange. It was not until I finished the thesis that I realized: the data exchange – even if it was perfect – is not what is needed! It just helps us to compensate the imperfectness of the CAx-systems and the workflow of product models. It took some time before I realized that what was really needed was *integration*, not data exchange. Starting an investigation of integration and its problems took a long time, and has involved a lot of modelling to overcome the complexity of the matter and to enable the search for a really generic solution.

Ever since I realized the importance of modelling I have tried to learn more about it to obtain more benefits from any area where it has to be applied. As it turned out that modelling itself is not extremely well studied, I decided to investigate it myself. This book is an attempt to systematize and make public all knowledge about modelling and its application in the field of engineering that I have acquired, together with my vision and as many ideas and small discoveries in the area as possible. I believe that once described, each good idea will sooner or later be understood, no matter how bad its description is. And if the idea finds the right public, it starts rolling and growing like a snowball down the hill.

Since I am neither mellifluous (most engineers are not), nor an English native speaker, it was clear to me from the beginning that this book cannot be – at least from a literary point of view – a masterpiece. But even knowing this, I thought that "seeding" ideas is much more important than achieving a high literary quality. I hope that many readers will not only understand and use the presented material, but will be able to explain, respectively describe it much better than me. Perhaps You will be one of those readers? Finally, as Francis Darwin supposedly said, "*in science the credit goes to the man who convinces the world, not to the man to whom the idea first occurs*".

In short, I hope that some of the ideas described in this book will either be useful to other people or lead to the birth of other novel ideas and thus contribute to the domain knowledge.

Topics/Keywords

Modelling, simulation, integration, reuse, lifespan, lifecycle, autonomy, intelligence, learning, complexity, efficiency can be mentioned as just a few of the more important topics.

Approach

I believe that the terms are tightly connected to the problems and their solutions (*cf.* Figure 0.1). In particular, the use of proper terms is very important not only to achieve the right understanding of the material, but also to avoid the emergence of pseudo-problems, misleading causes and side effects or improper solutions. Therefore, a great effort has been invested to define all terms used and to make the definitions as clear, precise and non-contradictory as possible. As such a goal is very difficult to achieve, some definitions can be unexpected or at least specific to this study.

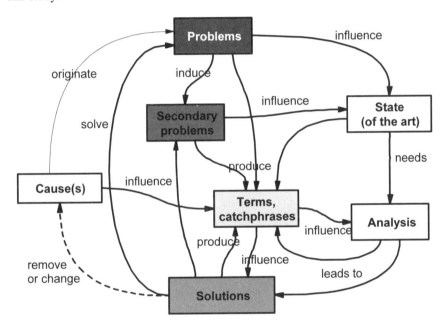

Figure 0.1. Interrelations between terminology and problem solving

A large number of earlier publications by the author are re-thought, improved and integrated in this text.

Ideas or concepts are sometimes presented on the principle of a "nasty commercial" – for some reason, unpleasant or annoying presentations can be memorized better and even against our will.

Some concepts, principles and approaches are used again and again in the reasoning, even when not apparent. This repetitive use happens not because of some particular preferences of the author (although preferences have definitely played a role), but due to their particular suitability for the task at hand. Two of them are the *set theory* and the *Pareto's principle*[1] (also known as the 80:20 rule), although their use cannot always be recognized at first glance.

---

[1] *cf.*, *e.g.*, Brockhaus, F. A. (Ed.) (1989) Brockhaus Lexikon, Mannheim, Deutscher Taschen Verlag

Audience

The issues discussed in this monograph are of interest to lecturers, researchers and students in the field of mechanical engineering, mechatronics or computer science, but the material could also be useful for programmers and other people interested in or practicing modelling of products or processes. I hope that the book can be of use also for people, involved in creating CAx-systems or dealing with them in any way – system architects, analysts, decision makers, *etc.* Finally, scientists dealing with either models or modelling in other scientific areas could find (parts of) the material useful.

Text Organization

The main text is organized in five chapters.

Chapter "Introduction" starts with the motivation for the writing of this monograph, discussed on the background of the actual problem area.

Chapter "Modelling Basics" defines basic terms in modelling, presents its objectives, and gives some possible modelling classifications based on different criteria.

Chapter "Conventional Product and Process Modelling" discusses in detail the problems of conventional product and process modelling, as well as some general problems related to the use of computer aided systems in different areas of mechanical engineering. Complexity, its consequences and the ways to master it, as well as the integration-related issues are just few of the viewed important topics.

Chapter "Towards Better Product and Process Modelling" starts with a preview of some of the known recent approaches in this area, which are trying to avoid the problems of the conventional approach. From the analysis of these approaches and from the modelling issues, reviewed in the previous chapter, is derived a set of requirements for an idealized "perfect" modelling approach. Later on, a novel approach to product and process modelling is presented, which exhibits potential for achieving better results with regard to reusability, integrability of heterogeneous models, flexibility, maintainability, *etc.*

The "Conclusion" and "Perspectives" present a final discussion, a general overview of future prospects as well as some plans for future work.

An index and a list of the used in the text abbreviations are included at the end of the monograph.

Disclaimer

As almost every other technical book, this one also does not contain only new material – in many places other people's views or opinions are presented, discussed or extended. The author did his best with referencing and giving credit always when possible. Should there be any occasion where this is not the case, it is not intentional. All trademarks and registered trademarks, mentioned in the text, belong to their respective owners.

Saarbrücken, March 2007   Nikolay Avgoustinov

Acknowledgments

This work would not have been possible without the infrastructural, scientific and moral support by the Head of the Institute of Production Engineering/CAM at Saarland University Professor Dr Helmut Bley, to whom I am very indebted. He was always the first who comprehended and valued most ideas of mine and encouraged their further development, even when they were far from obvious or not clearly expressed. His readiness to give a word of advice helped me very often.

I would also like to thank the Head of the Institute of Engineering Design/CAD at Saarland University Professor Dr Christian Weber for supporting me with software, hardware, know-how, sample geometric models and last but not least, with helpful discussions and advice.

Many thanks to all other colleagues from both institutes for their patience, cooperation and support. They have answered plenty of suspicious, sometimes unreasonable, and not always clear questions for years, always when asked - even in a foreign language or in bad German.

My deep gratitude goes to the system administrators of both institutes – Volker Henrich and Germano Porta – who always supported me with everything they could – hardware, software, literature, knowledge and advice.

Special thanks go to Chris May – a colleague and friend who was the first to suffer always when I needed help from an English native speaker.

I owe Renate Kröner special thanks for her support in all library-related activities and for all the photographs (no matter whether they appear in the book or not), which she always so patiently prepared for me.

I am deeply indebted to professor Dentcho Batanov for his moral support and valuable advice, as well as to my earlier colleagues Ilario Astinov, Dimitar Bojkov, Vassiliy Tchoumatchenko, Ivan Dantchev and many others who have been very often – even without knowing it – of great help.

I would like to thank all my (former) students who took part in the development of different models or have supported my work otherwise. Special thanks go to Stefan Gimmler, Jean-Claude Blummenau, Filip Bramowitcz and Tobias Geier.

Sincere thanks to Mogens Myrup Andreasen, Alex H. B. Duffy, Jürgen Gausemeier, Udo Lindemann, Joachim Lückel, Peter Nyhuis, Gunther Reinhart and Hans-Peter Wiendahl for giving me a permission to reproduce for the discussions pictures from their books, as well as to my colleagues Helmut Bley,

Marc Bossmann, Christina Franke, Lars Weyand, and Christian Zenner, who gave me their permissions to use portions of our cooperative works and publications.

I am very obliged to Springer London for the opportunity to publish this book. During its preparation I have got great support from Anthony Doyle, Kate Brown and Simon Rees. I would like to thank to all anonymous reviewers and especially to the copy editor, who improved the English of the whole book.

Many thanks also to the production editor Sorina Moosdorf and her colleagues from LE-TeX Jelonek, Schmidt & Vöckler GbR, who was of great help during the preparation of the book for typesetting.

Last but not least, I am enormously obliged to my family and especially to my wife for all the inspiring discussions, for the incessant proof-readings and decisive support as well as for their endless patience and understanding.

As is typical for an acknowledgement statement, there are many people that have direcly or indirectly contributed to the success of this work but are not mentioned in person here due to different reasons: to all these people I express my deep gratitude.

Contents

Figures

Tables

1

Introduction

1.1 Motivation

1.1.1 Challenges in Manufacturing, Products and Service Engineering

According to a document published by the Intelligent Manufacturing Systems Secretariat of the European Community, the following topics should be the focus of Manufacturing, Products and Service Engineering in 2010:

- Methodologies, tools, work environments for the conceptualisation, design, make of products and services/delivery, product support
- Integration of miniaturized devices and software into intelligent products
- Value creation processes in manufacturing (knowledge/information flow between suppliers and users) novel approaches to customization, logistics, maintenance
- Holistic product design/development and distribution tools and methods
- Global standardization initiatives: Inter-enterprise business processes (planning, scheduling, coordination); assuring process transparency, traceability of produced parts, shop floor automation/security
- "Knowledge communities" in production technologies, advances in virtual production, supply chain and lifecycle management, decision-aid systems, rapid manufacturing

Another publication defined a decade ago in Bollinger (1998) the Grand Challenges for the Manufacturing in 2020 as follows:

- Grand Challenge 1: Concurrent Manufacturing
- Grand Challenge 2: Integration of Human and Technical Resources
- Grand Challenge 3: Conversion of Information to Knowledge
- Grand Challenge 4: Environmental Compatibility
- Grand Challenge 5: Reconfigurable Enterprises
- Grand Challenge 6: Innovative Processes

Contemplating what is already achieved and what could be achieved in the near future, we have come to a vision that shapes the spirit of the present study.

1.1.2   A Vision: Manufacturing in the Twenty-x^{th} Century

Having a vision is not an end in itself. To have a vision means to analyse the past and the present and try to extrapolate development in the future in order to foresee it. The further away in the future one would like to (fore)see, the more of the past would have to be analysed and the higher the probability that the foresight will deviate from the reality when the considered time comes. Only those who (try to) foresee the future can anticipate the events – good or bad – and forestall the competition in dominating the market. To get the right vision is not easy, but even a bad vision is better than no vision at all. So let us try to recall the past of mechanical engineering, to review its present, to project ourselves into the twenty-x^{th} century of the millennium and have a look around. According to experience and imagination, each of us will be able to imagine a very different situation. For instance, at some point in the future the following could have happened:

I1.   Material processing has become infinitely easy, cheap and fast, which makes it affordable for everybody.

I2.   There is rarely a need for pre-fabricated materials, because people have mastered material transformation and improvement and can thus use the surrounding widespread materials or reuse materials and the energy of already unneeded equipment.

I3.   Due to I1 and I2, (conventional) factories are not needed anymore and are replaced by ubiquitous or home or in-place manufacturing.

I4.   Due to I1, I2 and I3, warehouses are almost not needed and the use of transport is significantly reduced, leading also to cheaper products and to reduction of pollution.

I5.   In such a situation, information and knowledge processing becomes the most important factor in manufacturing.

How realistic is such a vision? If it is realistic, when could we (or the future generations) expect it? In order to (try to) find answers to these questions, one has to consider the known achievements of the science and technology and attempt to estimate whether their further development can (at least theoretically) lead to similar results. No matter what conclusion is drawn, the reality can be different. Nevertheless, if we know what we would like to achieve, it is worth the effort to make the first steps towards the accomplishment of our desires as soon as possible, as well as to attempt to foresee and plan the remaining steps.

1.1.3   Preparing (the Technology) for the Twenty-x^{th} Century

No doubt, the technology of the next centuries will be better than the current technology. Nevertheless, despite different theories about giant acceleration (*cf.* http://accelerating.org   or   http://www.accelerationwatch.com/), technological singularity (http://www.singularitywatch.com/) and other predictions we do not believe in sudden or very rapid changes in the technology. Even if they happen, most of them usually have a limited influence on the technology as a whole and a strong impact within some particular field. We firmly believe, though, in the gradual but continuous improvement of science and technology altogether. For that reason, it makes sense to mention some of the promising contemporary technologies and to contribute to their development.

1.1.3.1 Thinkable/Conceivable Technologies

Rapid Prototyping (RP)
Rapid prototyping encompasses a number of technologies for fast creation of (real or physical) products on the basis of their 3D-models. The most popular until now RP-technologies have been selective laser sintering (known as SLS), stereo-lithography, 3D-printing and others. Objects are created as numerous thin layers of selectively hardened material having the necessary profile and created over one another.

The most often used materials are either powder-based (ceramic, metal or thermoplastics powder) or (photo-)polymer-based.

Intelligent Materials
According to Bullinger (2007, p. 36), intelligent (or smart) materials have the capability to react to stimuli from the environment or changes there and to adapt their functionality respectively. This is either directly possible or achieved through combining sensory materials with actuating materials and a control unit. The resulting combination is named *composite material* and has special properties.

Until now intelligent materials can be classified – according to the main effect they use or expose – in at least five groups:
- with shape memory;
- with piezoelectric effect;
- with electrostriction or magnetostriction;
- using electro-rheology and magneto-rheology;
- using chromogenic effect.

Speculating on further development, one could expect materials with programmable (or computer-controlled) behaviour (*e.g.*, remote form giving?) in the (near) future.

Better Use of Energy and Resources
Energy plays key role in industry, society and private life. Its ubiquitous availability, safety and price are factors having an enormous influence on its usability and on everything depending on energy. There seem to be at least three directions at the moment, offering very promising prospects – energy-saving technologies, use of regenerative energies (those of the sun, the wind, the water, the tides/waves, *etc.*) and recovery/reuse of energy (*e.g.*, from products to be recycled or from garbage). And if their development continues, it would not be surprising if in a couple of centuries (or decades?) people are in a position to draw all needed energy from the nearby environment.

Transportation
Transport is still one of the factors playing a major role in processes such as people transportation, delivery of raw material and goods, carriage/conveyance of (partly processed) parts during their manufacturing and many others. Despite all achievements in the area of transportation it is clear that there are many novelties

still to come. And for a great improvement it is not really necessary to have the transport technology of the spaceship Voyager – there are many already existing technologies that could bring significant improvement in the area even now – like the Levitation Railway or pipeline[2] transportation. The latter could be used, for instance, to easily deliver raw materials in fluid or powder form – *e.g.*, petroleum, paraffin, polymers, *etc.*, even to the ubiquitous home (nano) factories of the future.

Biotechnologies
Biotechnologies also have good prospects. Especially in combination with other sciences they offer attractive possibilities: biomaterials; material bio-transformations, bio-organization, and in the future possibly even genetically programmed material growth (up to programming of the final form!).

1.1.3.2 Human Resources and Human-related Technologies
Having good technology alone is not enough: we need experts who know how to handle it. In other words, no matter which technologies come to be used in the future, it is of strategic importance that their users can really control them. This means that, on the one hand, there should be no threats to humans or the environment, and on the other hand, the technologies should be used efficiently. A key factor for these two prerequisites is the *ability to understand* the technologies, which in turn requires the *proper qualification* of the immediately involved people. Such qualification is achieved by means of training and education, which can be enormously improved by use of appropriate models.

There are a number of novel learning technologies (eLearning, virtual and mixed reality, learning by playing, *etc.*), which have enormous potential. However, they can be used only with appropriate models or will bring more benefit when based on such models.

1.2   Immediate Goals and Working Areas

Assuming that we are motivated to achieve a goal, the next thing to do is to analyse the situation and to prepare a plan for the following steps. There is no plan proposed here, neither a detailed analysis. We simply try to share some observations and ideas for improvements that could contribute to conceiving the technology of the future. And the author's view at the time of writing is that our current position/advance on the developmental spiral requires first and foremost development and elaboration of concepts, methods, and tools for:

- information and knowledge representation
- conversion of information into knowledge
- automated decision making
- efficient product and process modelling, including reduction of the complexity of the modelling and the resulting consequences.

---

[2]   Not only for fluids, but also for small containers that can be used to carry objects in them.

Eventually, we need highly reusable and easily integrable models. And ideally they would be usable for *anybody*, at *anytime* and *anywhere*.

1.2.1   Information and Knowledge

Assume for a moment that the information and the knowledge are "made" of data. Data is a fascinating "material", which has one huge advantage – its *replication* and *transportation* are fast, easy and cheap. Information and knowledge inherit this advantage from the data, but they still have to be created, which is not trivial. But every technology is based upon knowledge, therefore, to prepare a better technology means that society has to extend its knowledge as quickly and thoroughly as possible. On the one hand, this means making it generally available, efficiently representable, easily reusable and understandable. On the other hand, all means supporting the knowledge elicitation, representation, processing, reuse, *etc.*, have to come (regularly) within the focus of the development.

This book is dedicated to modelling, as it is crucial for all aspects discussed above.

1.2.2   Elaboration of the Curricula of the Future

It would not be an exaggeration to say that the technology of the future – and therewith also the future itself – is forged now in the educational institutions, or more precisely – in the minds of the future generations. This means that we are responsible for keeping the curricula in these institutions appropriate and up-to-date, and for its gradual but permanent transformation into the curricula of the Twenty-xth century.

2

Modelling Basics

> *Science is build up of facts, as a house is build*
> *up of stones; but an accumulation of facts is no*
> *more science than a heap of stones is a house.*
>
> Henri Poincaré
> Science and Hypothesis, 1905

Before turning to the problems of modelling in mechatronics, which is our immediate domain of interest in this study, we need to set up a descriptive foundation by clarifying the essential notions, as well as to define new relevant terms, *i.e.* to give names to some (new) notions, when appropriate. A name cannot change the appearance or the properties of the named entity, but nevertheless, it has a great impact on people's attitude through associations that can be provoked – especially at the first contact with an entity. Improperly chosen or inadequate names can lead to misunderstandings and even misconceptions. As we shall see later on, names play an important role in communication, integration, standardization and other fields. We also use these names when we refer to the "building blocks of science" – notions, relations, facts, attributes and others. No "scientific house" (*i.e.* theory) can ever be built without such building blocks. For these reasons, special care should be taken when new terms are introduced or existing terms renamed. Due to the attempt to avoid invention of totally new words, the names chosen for some notions may seem strange at first glance, especially if considered out of context.

Now, let us begin with the formal meaning of the terms *model* and *modelling* and then discuss some of their most important attributes.

2.1 Models and Modelling

Let us start with the most frequently used terms and consider their motivation and interrelations.

2.1.1 Definitions

> *A definition is the enclosing a wilderness of idea within a wall of words.*
>
> Samuel Butler
> Note-Books

The word *model* is an overloaded term. For example, the Collins Cobuild Dictionary Sinclair *et al.* (1987), specifies fifteen meanings, with three of them – instances 1, 3 and 4 – being mostly relevant for our purposes:

> 1. *A model of an object is a physical representation that shows what it looks like or how it works. The model is often smaller than the object it represents.*
>
>> *...an architect's model of a wooden house.*
>> *...a working scale model of the whole Bay Area...*
>> *I made a model out of paper and glue.*
>> *Model is also an adjective.*
>> *I had made a model aeroplane.*
>>
>> *...a model railway.*
>
> 2. *...*
>
> 3. *A model of a system or process is a theoretical description that can help you understand how the system or process works, or how it might work. (TECHNICAL or FORMAL)*
>
> 4. *If someone such as a scientist models a system or process, they make an accurate theoretical description of it in order to understand or explain how it works. (TECHNICAL or FORMAL)*
>
> 5. *...*

Such overloading of the term with different meanings requires a clear initial statement of how we shall understand this term within this study. Let us have a look at some more specialized (*i.e.* not so universal) definitions.

In Stachowiak (1973), any object having the following three main distinctive features is viewed as a model: to be a representation of something, to be a simplification and to be pragmatic (in the German original they are called "Abbildungsmerkmal", "Verkürzungsmerkmal" and "Pragmatisches Merkmal", respectively). Actually, the first one seems to be not always required – *cf.* Section 2.1.3.2 below.

Yet another definition is found in Woolfson and Pert (1999):

> *The essence of the model is that it should be a simplified representation of some real object or physical situation which serves a particular, and perhaps limited, purpose.*

Although these two definitions might seem different at first glance, what in the latter definition is expressed as "to serve a particular purpose" is formulated in the former definition as "to be pragmatic".

In essence, each model is a purpose-dependent representation. According to the purpose of modelling, it might be required that different traits are represented or respectively ignored, therefore, we shall define model as follows:

Definition 2.1: *Model is a purpose-dependent, finite, simplified, but still adequate representation of whatever is modelled, allowing us to abstract from its unimportant properties and details and to concentrate only on the most specific and most important traits.*

The respective implementation may use different media or principles (*cf.* Section 2.4.2.3.1 below) and is neither substantial nor pre-defined. When the model is a *representation of the object's traits and their interrelations* by means of pieces of information (or *data*), we shall speak about *informational models* or *data models*. When these pieces of information are *electronically representable values* (numbers), we shall speak – depending on the context – about *software models* or *computer models*.

To better understand the nature of models and modelling, we shall first of all examine how these notions are related, and how they depend on other factors. A somewhat humorous interpretation of what we are concerned with here is sketched in Figure 2.1.

Modelling

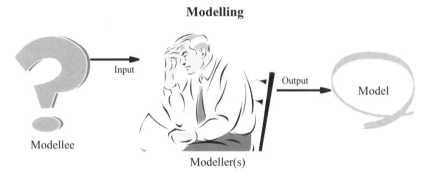

Figure 2.1. Modelling: the "holy" trinity

In a typical representation of a process, the assumption is made that somebody or something acts on something else (input) to create or achieve a result (output). Apparently, such a scenario would not represent the really important fact that processes are time-dependent, *i.e.* they "progress" with time. More precisely, it is the process of creation and development of models that is understood as *modelling*. No unambiguous generic (preferably: one-word) term exists for what occurs as input of the modelling process. Expressions like "object, product or process to-be-modelled", for instance, would be rather long and imprecise. As we have to refer to the input of the modelling fairly often, let us consider introducing a new term instead. The latter would be used as a generic term in all cases of modelling and especially in the lifecycles of both original and derived products (*cf.* Section 2.2.3). Having looked into the existing terminology as well as having considered the possibilities of making up a new, "artificial" term that would simultaneously be intuitive, short, well-known and, at the same time, not contradictory, I've come to the conclusion that the word *modellee* will be most appropriate for my purposes. This term has the same root as model and modelling and in addition makes allusion to words with similar morphology and well-known meaning like employee (person who is employed), trainee (person who is trained), adoptee (person who is adopted) and many others. So we shall use the term modellee for referring to *what* is or will be modelled (the object of modelling).

A conceivable approach to defining the term *model* is to enumerate whatever appears to be related to any thinkable model, and then show the relation between these enumerated "components" or attributes. A graphical representation of such an attempt is given in Figure 2.2. Since for the preparation of this picture we have abstracted from insignificant properties, concentrating only on the specific and important ones, we have eventually created a *model of a model*, or a *meta-model*.

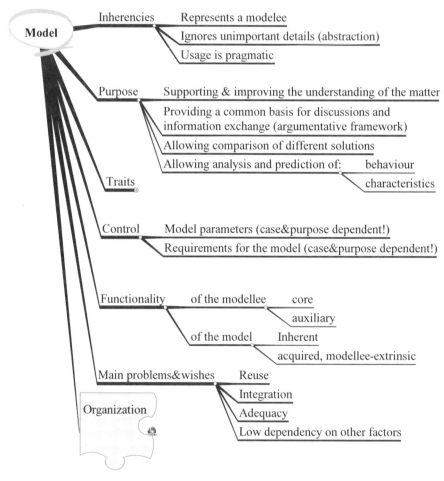

Figure 2.2. Attributes of a model and their relations

Of course, not everybody will accept the representation given in Figure 2.2 without objections. But, depending on the purpose for which this meta-model is created, different requirements are imposed on its representation, behaviour, level of detail and so on. Inasmuch as these requirements are case-dependent, no model can be perfect *per se* (or in general, or for all cases), but any model can be perfect for a given purpose. And the purpose of the meta-model in Figure 2.2 is to give us an idea what the main attributes of every model are, and how one could start to develop a model. In order to be able to show it in more detail, the organization of a model is presented in a separate picture – Figure 2.3.

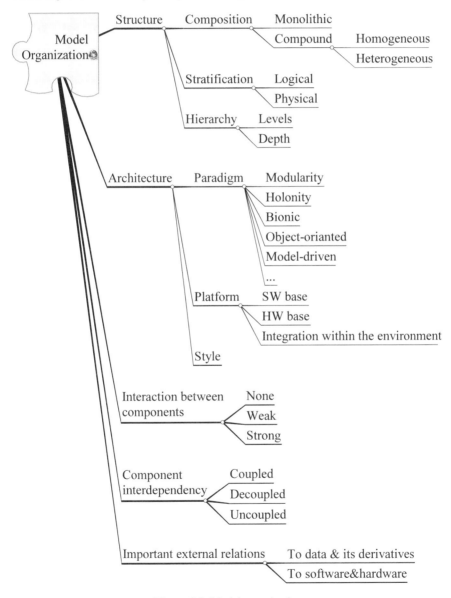

Figure 2.3. Model organization

Another more detailed representation of the most important participants of the modelling process and the relations among them is given in Figure 2.4. Such representation has some similarities with the conceptual graphs as they are defined in Sowa (2000, Appendix A), but the conceptual graphs possess greater expressive power.

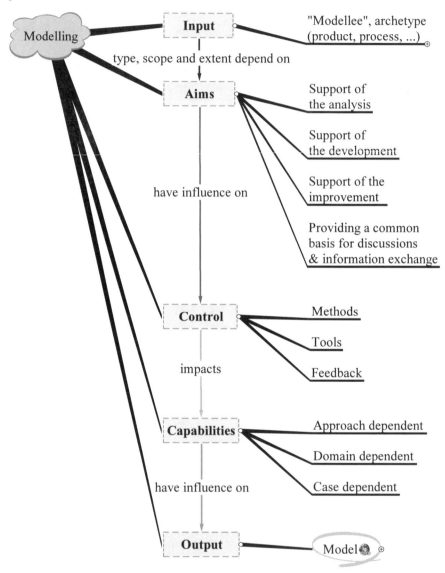

Figure 2.4. Participants in the modelling process and their interrelations

Quite in the spirit of the model of a model given in Figure 2.2, we can prepare a model of the modelling process itself (*cf.* Figure 2.6), which will enable its systematic study. As we can see there, the result of the modelling process is a (possibly compound) model. If the content of this monograph is brought in relation to the specificities of the modelling process illustrated in Figure 2.6, we can say that it focuses on information-technology aspects and approaches for efficient, platform-independent modelling in the area of mechatronics and mechanical engineering, based on an arbitrary web-browser, a (formal) modelling language, and a 3D visualization engine.

2.1.2   Modelling Stages

Modelling is a complex iterative process and has typically several *phases* or *stages*. Consider the following quotation from Duffy and Andreasen (1995):

> *Phenomena models are primarily based upon observations and analysis of the "reality" of design and the use of the tools employed, and hence reflect "descriptive" models. Where appropriate, these models are then developed in more detail as information models and similarly as computational models and tools. At each stage any model can be compared or evaluated against any previous model in order to enhance our understanding and hence models.*

According to the graphical representation in Figure 2.5 based on Duffy and Andreasen (1995), the nodes in the bottom row of the figure can be viewed as modelling stages, with the reality being the origin of (computer) modelling, and the development leading from a phenomenon model through an information model towards a computer model.

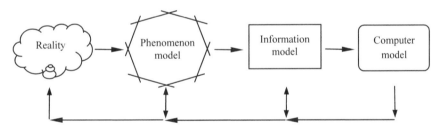

Figure 2.5. Design modelling research approach, after Duffy and Andreasen (1995)

At a careful observation, many (nested) cycles can be discovered in Figure 2.5, and it is not easy to tell where the "beginning" is meant to be: when we have to do with cycles, the cycle-start can be everywhere. Note, however that two main types of activities exist during the development of any model:

1. *Essential (modelling) activities*: improving the model and its adequacy by increasing the number of modelled properties, their accuracy and other essential qualities; and

2. *Auxiliary activities*: "fighting" with the restrictions, limits and problems of the modelling approach used, methods, tools, *etc.*

Thus, at each new stage some new model quality is achieved, but at the price of increased auxiliary activities. The *efficiency* of the modelling is directly proportional to essential activities and inversely proportional to auxiliary activities.

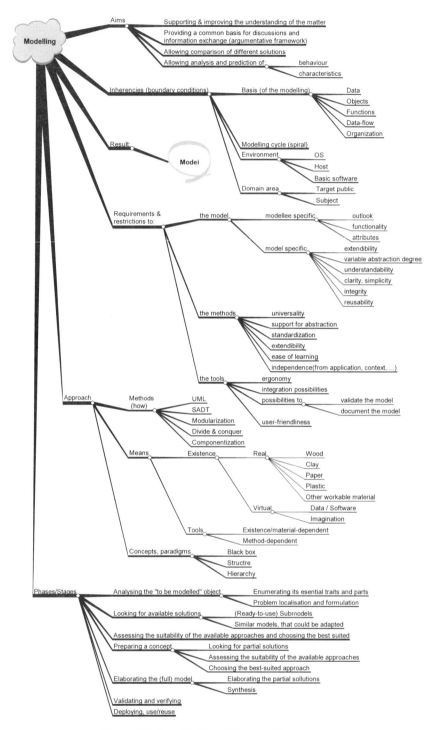

Figure 2.6. Specificities of the modelling process

The stages in Figure 2.5 are defined on the basis of "model metamorphosis" during model development. If, instead, the modeller's activities were considered, it would certainly be possible during the model development to distinguish phases similar to those presented in Figure 2.7.

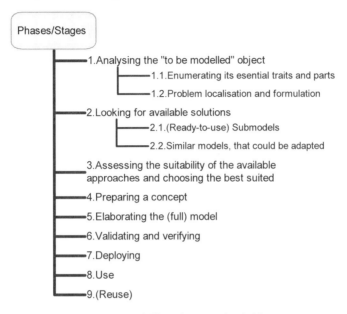

Figure 2.7. Modelling phases and activities

2.1.3   Purpose and Objectives of Modelling

Ipsa Scientia Potestas Est
(Knowledge is power)

Francis Bacon, Religious Meditations, Of Heresies

It is in the nature of mankind to strive for power in the most general possible sense – the ability to rule, govern or control every possible thing from the simple toy to the whole universe. For people of different professions this general "rule" sounds somewhat different – the politician strives for political power, the physician would like to have power over all diseases, the fireman wishes to be able to control each fire, the engineers pursue control over the production and so on. But the sense remains in all cases the same: to hold full control over the respective domain. Unfortunately, there are powers that we cannot (yet) control – like the four elements, the sun, the cosmic powers. In cases where uncontrollable power (often referred to as *force majeure*) is involved, the next most attractive and important option is the ability to predict the flow of the upcoming events and the near future. For instance, science and technology are not (yet) strong enough to prevent an earthquake or tsunami-waves, but the foreseeing of their oncoming, together with the appropriate actions, can avoid almost as many casualties as its prevention and thereby avert calamity.

Let us introduce definitions of the terms *control* and *prediction*, which will be adequate for engineering purposes.

Definition 2.2:   *To predict means to know the causal connection between some event e_0 and its consequence c_0. We shall call e_0 a* forerunner *of c_0.*

Definition 2.3:   *To (fully) control a given object or system means to be able to put the system in any of its known states, whichever of them is desired.*

In many cases Definition 2.3 does not hold, but it is possible to avoid certain undesired states of the system. Such an ability is more important than it seems at first glance. For instance, the ability not to allow a system to reach its worst (or most dangerous) state is more important than the ability to switch this system from the worst state into any other state. We shall call the former ability *blocking* (or *weak*) *control* and define it as follows:

Definition 2.4:   Blocking control *is the ability to recognize a forerunner and issue a reaction r_0 to it in a way that avoids an (upcoming) undesired event or state of the system, as well as any undesired consequence.*

Since, according to this definition, the controller first waits for a given forerunner to occur and then issues a reaction, we shall call this kind of control *passive control*. Of course, in many cases the controller can act on its own – *i.e.* without waiting for any forerunner – in order to change the state of the system. We say then that the controller is *proactive* or exercises *active control* over it. The actions or reactions, issued for gaining or keeping control over a certain system or object are usually called *commands*.

Both passive and active control can be possible either for all forerunners or for some of them, so we can speak about full or partial control.

Definition 2.5:   *When for a given system and a person controlling it Definition 2.3 holds for any forerunner, this system is fully controllable by the mentioned person.*

Cases where a system is fully controllable by somebody are rather rare. Therefore, when we speak about control we typically understand a highly, but still partially controllable system.

The ability to predict, exactly as the ability to rule or control, can exist on any level of scale between the macro-cosmic and the micro-cosmic level. Somewhere in the middle there is also a level that can be viewed as the engineering level, which is in the focus of this work. But is the modelling really related to power and prediction? If yes, how are they related? Well, neither controlling nor prediction are possible without the appropriate *knowledge*. The approximate interdependence of these two abilities on *a priori* knowledge is illustrated in Figure 2.8.

So, the next question is how the knowledge about a given topic or domain is acquired and whether we could influence the acquisition speed. Let us discuss these topics with the help of Figure 2.9. Suppose that the abscissa in Figure 2.9

shows the time or, in some cases – the lifetime of a given domain or entity. Then the solid line, starting from the beginning of the coordinate system shows how our knowledge about this domain or entity changes with the time. Four different phases are denoted under the abscissa, through which the "dealing" with knowledge goes – passive and active learning, and passive and active use of knowledge. In addition, there are two areas surrounded with rectangles, whose line patterns are different. Each of these areas is annotated with text, describing the (specific) activity, which is possible only with an amount of knowledge greater than that corresponding to the knowledge curve at the lower left corner of the respective area.

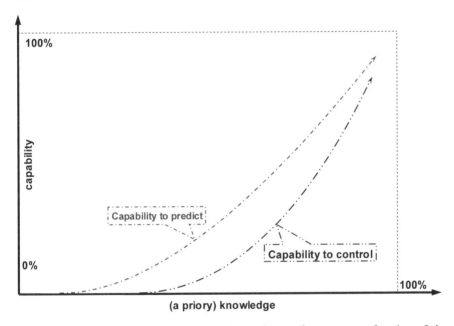

Figure 2.8. Domain-related abilities to predict and control events as a function of the acquired domain-knowledge

During the first phase (passive learning) knowledge is collected relatively slowly, mainly through *perception* – an activity that takes place during all four phases and thus exists throughout the whole lifetime; it is marked with a rectangle, bordered by a dot-pattern line. Perception can occur through any of the human's five senses, but most of it happens through *observation*.

Observation helps us learn everything that can be seen, but since most of the objects are opaque, neither their structure nor the connections among their components/elements can be studied through observation. Imagine you are presented with an unknown to you until now electric torch, having no battery yet. How do you know which position of the switch turns the light on and which turns it off if there is no inscription? One possible way to tell is to put a battery in the torch and to try which position turns the light on, but this is already another activity – an *experiment*.

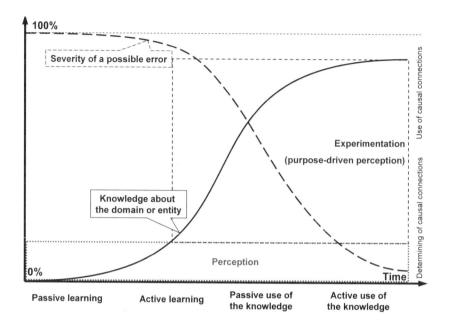

Figure 2.9. Knowledge acquisition and use

Having collected enough *basic knowledge*, an (intelligent) individual is capable of conducting some experiments, observing simultaneously the causal connections between his own *actions* and the subsequent *reactions* of the entity, increasing thus additionally his knowledge. In the case of the electric torch the basic knowledge is that it can light that it works with a battery, and that the light can be turned on and off through the switch. The action would be turning the switch and the reaction – changing the state of the light from on to off or *vice versa*. As soon as you can connect each position of the switch with a state – either lighting or non-lighting – you have learned to use the switch (and the electric torch). Now assume that a friend of yours has been with you and has watched attentively your actions all the time. Would he also have acquired the same knowledge as you did? Well, almost: he would know which position of the switch turns the light on, but he would not know, for instance, how much force is needed for turning the switch. The point is that your friend's newly acquired knowledge is gained through *passive experimenting*, while you gained the same knowledge through *active experimenting*. In general, more knowledge can be gained through active than through passive experimenting.

After sufficient experimenting, the acquired knowledge reaches another point at which the individual is able to *control* the entity to some extent (*cf.* Figure 2.8 again, with the domain being the use of an electric torch), so that some immediate goals can be reached – *e.g.*, turning the light on or off, changing batteries, *etc.* The acquired knowledge typically does not increase during controlling; instead, the *a priori* knowledge (*e.g.*, from old experiments) is confirmed and increases thus the certainty of the controller that the known commands can put the controlled entity to certain states or prevent it from getting into undesired states. Only if something unknown happens – say (if we continue our mental experiment with the above-mentioned electric torch), the light suddenly does not go on anymore due to a burn-out of the light bulb – this can be registered as a new possible state and thus increases marginally the domain knowledge. The knowledge how to return the system from an unknown state to a known one can be either still missing or generally available – *e.g.*, pressing the reset button of any computer puts it into a well-known state. Thus, the controlling itself can lead to acquiring new knowledge only indirectly – through coming to unknown problems or situations, solvable (only) by means of experimenting. The last proposition raises a fundamental question: is it possible to gain knowledge by mental experimenting? A positive answer to this question would revolutionize the whole science by making many experiments needless. What would be possible for sure is to search the memory for patterns of similar but already solved problems, and then attempt to derive from any pattern found a solution for the current problem/task.

Since in this phase the main activities are observing and experimenting activities together with some controlling, an appropriate name for the phase is *passive use of knowledge*.

From another point of view, as soon as an individual has noticed the causal connection between two (types of) events he can predict what would happen after some known forerunners occur. Back to the example with the electric torch: after using it long enough one should know that a noticeable decrease in the light intensity means an approaching end of the battery's charge, *i.e.* it is possible to *predict* the need for a replacement in the near future, and to take care to have the battery ready at hand. The prediction itself does not increase the knowledge, but as soon as it becomes clear whether the prediction is true or false, the knowledge very often may increase[3].

The most important outcome of the phases passive learning, active learning and passive use of knowledge is the determining a causal connection in the given domain. After acquiring a reasonable amount of causal connections (50–80%), it becomes more and more important to use them for acquisition of additional knowledge. We shall call such activity an *active use of knowledge* and name after it the last phase in the learning process. The reasoning and more specifically the use of techniques like induction, reduction and deduction for acquiring new knowledge on the basis of available knowledge and information are typical examples of active use of knowledge.

Let us summarize again the analysis of Figure 2.9:

- Apparently, the activity leading to the highest learning speed is the "experimenting". Clearly, well planned experiments can additionally increase the learning speed.
- It is never possible to get 100% of the theoretically possible (or practically available) knowledge because it is impossible to learn everything through observation and experiments. Instead, the curve showing knowledge acquisition as a function of time provokes associations with the Pareto-

---

3   It depends on the kind of prediction, though. If you predict that the next toss up would be a head, you would not know more after the coin falls tail. But if you predict, say, that increasing the pressure in the tyres of your car twice would prolong their life, and yet the first tyre explodes during the increasing, you would know more as a results of this experiment.

principle: in many cases it is possible to acquire about 80% of the knowledge about a certain domain in about 20% of the time.

- The probability of each prediction coming true is proportional to the knowledge about the respective domain.
- It is impossible to fully control anything that is not well-known (or learned). On the other side, only tasks small enough to allow knowing everything about them, can be automated. Assume the control (of a process) can be defined as a mapping of a set of input states – problems – to a set of output states – solutions. Since the automation can be defined as delegating the control or the decision taking to an artefact (device, software, combination thereof, or whatever else), the existence of such a mapping and the possibility of its implementation are crucial. The implementation is only possible if: a) the number of probable input states is countable and exactly known, and b) an onto-mapping M of the set of problems $\{P\}$ to the set of solutions $\{S\}$ is known, and c) M is realizable as an artefact.

So, how could models help here? At least the following two reasons for using models are justified:

a) Models can be used instead of real resources, at least during the early phases of the development, and thus make even the most intensive experimenting affordable and (financially) more effective.

b) Models can save time when they are workable or when they allow automation of experimenting. For software models both conditions are fulfilled.

In short, the objective of modelling is to increase learning speed and the amount of acquired knowledge (reason b) and simultaneously decrease the costs of knowledge acquisition (reason a), supporting thereby indirectly the abilities to predict and to control. On this basis are built concepts like Digital Factory, Virtual Factory, or Smart Factory: if anything that has to be build in reality – from a given product up to the factory producing it – is fully modelled, studied and optimized in advance, there is a great potential for saving time, money and other resources.

Of course, there are other reasons why we need models, which are more or less directly related to the discussed abilities to rule and to predict. They will be discussed in the following section.

2.1.3.1 Why are Models Needed

There are so many reasons for using models that their complete enumeration and description is almost impossible. Nevertheless, let us try to consider some of the more important ones (*cf.* Figure 2.10).

Models contain or reflect only the most important, for a given purpose, traits of whatever is being modelled. As a result, they reduce the complexity of the modellee and allow the modeller to ignore unimportant traits in order to concentrate on the essentials. Therefore, models crucially support and improve the understanding of the matter. Since the models are a simplified, finite representation of something, they are easier to handle. In many cases the only way for comparison of different objects, products, solutions, *etc.* is to compare their models. For instance, we (still) cannot compare two screws atom-by-atom, particle-by-particle, and this would not make sense either. But it does make sense to compare their diameters, lengths, pitches, number of threads and a couple of other purpose-dependent traits. Since these traits represent a kind of screw-model, it is enough to compare the models instead of the modellees.

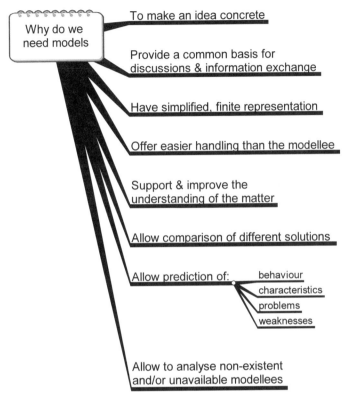

Figure 2.10. Some reasons to use models

Another interesting application of models is the prediction of properties and behaviour. This prediction is based on the comparison of relevant characteristics with those of similar but already known objects, activities, *etc.* For example, whenever we see that a bolt and a nut have the same diameter, pitch and number of threads, it is possible to predict on the basis of previous experience that the bolt will fit into the nut.

A careful look at Figure 2.11 reveals that the model is involved in two loops: gaining insight and applying it to the problem in order to solve it. Besides, the former loop is nested in the latter. The advantage of the "long way" from the real system through problem definition, target definition, model creation, experiment, analysis and so on, viewing the steps clockwise, is that the inner loop (gaining insight) allows one to collect the knowledge, necessary for the solving of the problem, much more quickly (*cf.* Figure 2.9 again!). Note that the more iterations are made in the inner loop the deeper insight would be gained, and the problem should be solved either sooner or better, or both.

Another important reason for using models is to harness them in the problem solving process. A nice example of how this could be done is given in Nyhuis and Wiendahl (2004) and reproduced in Figure 2.11.

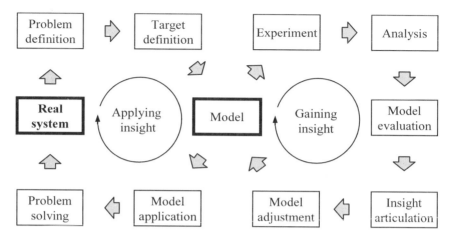

Figure 2.11. Model-based problem solving process, after Nyhuis and Wiendahl (2004)

2.1.3.2 How Models Arise

Some of the possible ways for creating a model are represented in Figure 2.12.

In all cases, after a model has been created, it has to be represented in some way in order to make it useful. Without such a representation, the model creator cannot communicate the model to other people, which renders it hardly usable.

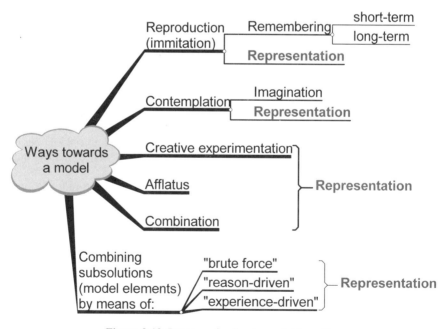

Figure 2.12. Processes leading to model inception

2.1.3.3 Models and Product Development

It seems at first glance that development is a linear or straightforward process, but this is not really the case. The point is that the real flow of this process is difficult to represent. Indeed, attempts to represent development typically concentrate on the most important traits of the process and ignore the non-essential ones. In particular, models are indispensable for the product development, and the main reasons for using models can be summarized as follows:

1. to support the decision taking
2. to shorten the development time
3. to minimize the development costs

All three reasons are more or less interdependent (at least in the direction from reason 1 towards reason 3 – *i.e.* easier and faster decision taking can shorten the development time, which in turn reduces the costs.

A compact explanation of reason 3 is given in Figure 2.13. The three curves reflect the nearing of the *product's achieved properties* to the *requested properties* of a product as the development advances. The right-hand solid-line curve refers to the normally developed and produced artefacts; the left-hand solid-line curve refers to a rapid prototyping and the dashed-line curve refers to a virtual prototyping – *i.e.*, to the percentage of *modelled product's properties*.

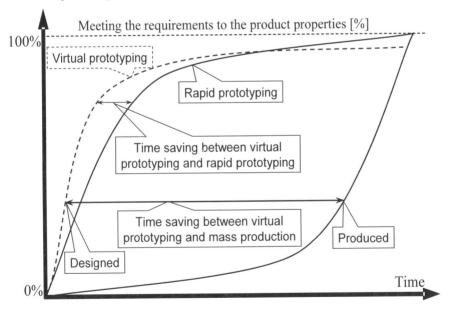

Figure 2.13. Maturity of model and (real) product during their development, after Gausemeier *et al.* (2000)

Why are these curves so different? There are several reasons for this. On the one hand, before an artefact can be produced, the respective production process has to be developed and implemented. On the other hand, for the production of the rapid prototype another technology is used, which leads quickly to a product, but is much more expensive. Due to specificities of the rapid prototyping technology in some cases the manufactured prototype does not have all properties or does not have the same quality as the real artefact. This applies to the virtual prototypes (or virtual models) in an even stronger way: since they are immaterial, they always lack more properties than a prototype. Since the model is always simpler (by definition) than the modellee, its processing is also easier and much faster. Moreover, since development is a cyclic process, the time difference gained on each loop accumulates. Furthermore, the development of models (and especially software models) requires fewer resources than the development of the product itself or than the manufacturing of its (rapid) prototype, and can be therefore much cheaper. Consequently, it is affordable to make new iterations of the development cycle even with minimal corrections of the model, and the development progress is faster.

Use of models is indispensable also in situations when products or processes are developed for still unknown application areas – *e.g.* spaceflights.

In short, a careful look around confirms once again that the whole science is based on models. Without models it would be impossible to analyse, to communicate, to compare, to take decisions, to improve, to solve problems and so on. The question about the form of existence (*cf.* Figure 2.2) of a model is of a lower importance, as long as the model achieves its purpose. Accordingly, the capability to control the modelling (of everything and everywhere) can also be crucial for success.

Figure 2.14. Important milestones of the product lifecycle and their sequence

2.1.3.4 Models of Product Development

Since the shortening of the (product) development cycle and the improvement of product quality can be viewed as main purposes of the modelling, it is instructive to discuss this development on the basis of a simplified model. The first question is when the development of a given product starts. Assume that there exists increasing demand for a given product. After this demand reaches a certain threshold (*i.e.* after some delay), a formulation of requirements for the product begins. As soon as the requirements are considered complete, the (actual) development can start. With the development progress, more and more of the functionality of the product is finished, which means that more and more requirements are satisfied. Again, after reaching some threshold of functionality, mass production can start, followed – with respective delays – by marketing and use of the product. With the start of the product use starts also the wear and tear of its instances, and during use, new requirements are posed. Thus, the loop is closed, and we can speak about the lifecycle of the product with its elements or milestones. A simple model, illustrating the sequence of the mentioned milestones is sketched in Figure 2.14.

Some of these milestones are in antagonistic relations, *e.g.*, demand *vs.* supply, or unsatisfied requirements *vs.* achieved (required) functionality. The latter is more interesting from a technical point of view: as soon as the formulation of requirements is complete, the development starts, and with the fulfilment of each required function or quality the number of unsatisfied requirements decreases until all of them are satisfied. At this point the number of unsatisfied requirements is 0 and the required functionality is 100%, and the first development cycle ends here. At the latest with the beginning of the product use, though, new requirements can arise, which may cause the next loop of the cycle to start. A simplified model of this process is illustrated in Figure 2.15.

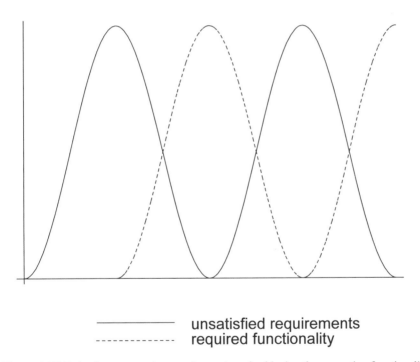

unsatisfied requirements
required functionality

Figure 2.15. Delay between posing requirements and achieving the respective functionality

Typically the number of requirements decreases exponentially with every new loop. This fact can be reflected in the model as illustrated in Figure 2.16.

If we define *model maturity* as the difference between the unsatisfied requirements and (implemented) required functionality, the resulting graphical representation of the curve will be very similar to the upper curve in Figure 2.13.

2.1.4   Some (Unusual) Examples of Models

Let us view some examples and see how they comply with the definitions, and decide which of them are models and which are not. Paradoxically, these samples show that people are modelling all the time, even if they are not aware of it.

2.1.4.1 Text

Strangely enough, the most often used models are *text models*. Let us first consider the single words: they are finite, simple and are "linked to" or describe more or less adequately something. For instance, any verb describes a process or an action; the verb alone is hardly sufficient to achieve an accurate representation, but it represents the most important trait of the action. Most nouns cause association with specific objects or even classes of similar objects – *e.g.*, the noun "lathe" is normally associated with the respective machine tool. When somebody says, "I am running", we can imagine that he is moving in such a manner that his body periodically has no contact with the ground. Thus, on the one hand it seems that most words can hardly be viewed as models. On the other hand, every word is related to some concept or notion in our minds, which is in turn a simplification or idealization of something and can, therefore, be viewed as the model of this something.

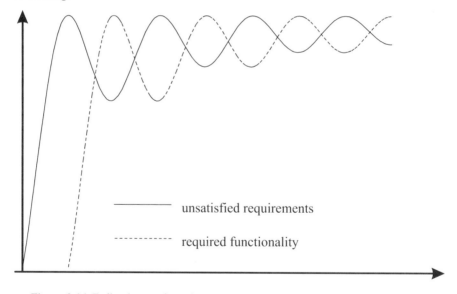

unsatisfied requirements

required functionality

Figure 2.16. Fading in posed requirements and achieving the respective functionality

Now let us consider an arbitrary (textual) description of something. No matter how detailed the description is, it could not describe every atom, every bit, every detail – simply everything – since the description would become infinite. Thus, the author of the description tries to describe the essential things first, then some less important and so on, until there is no reason to describe further (levels of) details. Two criteria are most often applied in deciding when it would make sense to stop, namely:

a) it is clear that the description of more details would not contribute to the purpose for which the model is created; and

b) the effort for describing more details would be greater than the achieved benefit.

Thus, the description is a kind of representation of an object, it would become finite due to the outlined reasons, *i.e.* by applying one of the above mentioned criteria or both of them, and it would contain the most important traits of the modellee by ignoring whatever appears to be insignificant. Summing up, words can be viewed as models of notions, texts as models of ideas.

2.1.4.2 Drawings, Sketches and Maps

Similar reasoning is applicable to drawings, sketches and maps: they are finite and represent only traits that are important for a given purpose. Nevertheless, not every sketch is a model: the pen scratches, made by someone unconsciously – for instance, during a phone call – would seldom be named a model of anything.

Maps[4] of the same landscape or area made in different scales offer us a remarkable example of models with different levels of detail. An indispensable element of almost any map is the legend, where a correspondence between real objects and their representations on the map is defined.

2.1.4.3 Pictures

Pictures in the form of photographs, drawings, scans, *etc.*, are also finite and simplified representations of something. Since they represent the outlook of something, they can be viewed as models of the respective outlook. In many cases, though, the outlook is not the most important property; in such cases pictures can hardly be called models. For instance, a photograph of a book can be a model of the appearance or of the design of a book, but not of the book itself.

It is interesting to contemplate whether a photograph of the content of a book, with still readable text, can be viewed as a model of the book. At first glance, it is a finite representation of the content of the book. At a more careful viewing, it becomes clear that the real model of the book is the text of its content, while the picture is rather a representation of the model of the book and, in a sense, it can be viewed as a meta-model of the book.

2.1.4.4 Bank Statements

A bank statement is a textual document representing the financial transactions of a customer for a period of time (usually month or year). The question is if this statement, which is final and represents important milestones of the customer's financial or bank-related behaviour, could be viewed as a model of this behaviour. The answer can be figured out if the type of information available in such statements (amount transferred, date, source, destination) is compared with the "parameters" of financial behaviour – *i.e.* frequency of transactions; volume of month's and year's turnover; minimal, maximal and average value of transactions, *etc.* It is clear that these types of information are different. Therefore, although a bank statement can be used to organize a model of someone's financial behaviour, it cannot be viewed as its model.

---

[4] We mean here geographical, geological and similar types of maps.

2.1.4.5   Itemized Phone Bills

For similar reasons an itemized phone bill is not itself a model of a person's communicational behaviour, but rather could be used for building such a model.

2.1.4.6   Model of a Circle

The geometric figure circle is so well-known that having heard or read this word everybody can immediately imagine its specific shape. Therefore, if we are discussing geometric objects where the respective shape is important, the word "circle" can be viewed as a model of this class of objects, since it bears the most prominent trait of the class – its shape.

In the majority of cases, though, we would like both to use the same model for the whole class and to be able to represent different specific objects (instances) of the same class without having to model them again. To achieve this we need to factor out everything that is specific for the whole class and make some characteristics serve as parameters, allowing us to instantiate arbitrary representatives of the class. Then, in order to be able to distinguish between different objects of the same class we need only to compare the values of the respective parameters.

In the case of a (two-dimensional) circle we would need at least the radius, which can be represented by means of one number. Should we need to distinguish also between circles with equal radii, we could use in addition the coordinates of the circles' centre, which are a pair of numbers for each circle. Consider the hierarchical representation in Figure 2.17. The radius and the centre point are grouped and represented as properties. Further specifications account for model-specific names, types of properties and validity rules (denoted as constraints).

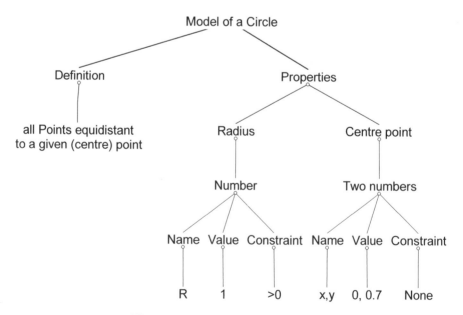

Figure 2.17. Simplified model of a circle

An interesting point for discussion could be whether the definition itself is a property or not. Later on we shall discuss other possibilities, but for the purposes of the current simplified model, let us leave it the way it is now.

2.2   Model-related Terms and Notions

> *The task of formalizing everything is like the construction of a medieval cathedral: it takes centuries to complete, and when it is done, someone else will have a plan for an even grander cathedral.*
>
> John F. Sowa
> Knowledge Representation

Numerous terms are used in the literature in connection with models and modelling. Some of them are already well known and established, some, still contradictory or simply not popular. In order to avoid misunderstanding, we introduce in this section the terminology to be used throughout this monograph as well as some characteristics and properties of models.

2.2.1   Prototype

In mechanical engineering, we often hear the term *prototype* in addition to the terms model and modelling. The definition in Sinclair *et al.* (1987), for example, puts prototypes in a strange relation to models, namely, prototypes seem to be restricted to models only, which is certainly not the case. "*A prototype is the first model that is made of something; P. is used as a basis for later improved models*". In yet another dictionary, we read: "*A prototype is a new type of machine or device which is not yet ready to be made in large numbers and sold.*" And as a second meaning: "*if you say that someone or something is a prototype of a type of person or thing, you mean that they are the first or most typical one of that type.*" Obviously, the latter two interpretations are more precise and do not lead to contradictions. If we replace the word "model" with the word "instance" in the first definition, all three interpretations would become compatible. With this adjustment, we can adopt the following definition for the purposes of the current study:

Definition 2.6:   *A prototype is the first instance that is made of something; P. is used as a basis for later improved instances.*

The prototype is typically a real material thing, but in some cases it can also be virtual – in the sense of *imaginary* or not perceivable by the five human senses. The possible combinations between the type of the prototype, the type of the mature product (or end product) and their "virtuality" are sketched in Table 2.1. We can see in this table that the strangest predicted combination would be to prepare a real (in the sense of material) prototype of a virtual end product.

A similar comparison of modellee types with prepared models demonstrates that in this case all predicted combinations are possible (*cf.* Table 2.2).

Within any product's lifecycle the prototype comes clearly before the mature product. But on the basis of what is the "first instance" made then? And if we can model real objects, is it possible to model objects that still do not exist? And how should they be called?

Table 2.1. "Virtuality" of prototypes and models

#	Mature product	Prototype	Possibility	Plausibility	Example
1	real	real	yes	yes (most common case)	mock-up
2	real	virtual	yes	yes	digital mock-up
3	virtual (unperceivable)	real[5]	hardly	?	?
4	virtual	virtual	yes	yes	software

Table 2.2. "Virtuality" of modellees and models

#	Modellee	Model	Possibility	Plausibility	Example
1	real	real	yes	yes (most common case)	mock-up
2	real	virtual	yes	yes	digital mock-up
3	virtual (unperceivable)	real[5]	yes	yes	sketch of a magnetic field? listing of a program?
4	virtual	virtual	yes	yes	software

2.2.2   Archetype

It is possible to model everything – modellees can be existing and non-existing, real or virtual, abstract or concrete. Often the possibility to distinguish between the modelling of already existing and modelling of not yet existing modellees is crucial. The latter case is of special interest, since it is specific for every novel product or process. The models in such cases are successors of ideas, but in our view there is one additional intermediate stage between a new idea and the model or prototype of any future product: the *archetype*.

In Sinclair *et al.* (1987) the term is explained as follows: "*An archetype is something that is considered to be a perfect or typical example of a particular kind of person or thing, because it has all their most important characteristics.*"

The archetype can be viewed as a mature well-elaborated idea, which can create a clear vision of the modellee in a modeller's head. It is a bearer of the

---

[5] In this case "real" is used in the sense of "made of some material".

inherent (or most important) traits and functions of the desired (or designed, or modelled) object, product or process.

Definition 2.7:   The elaborated idea that is (or has to be) modelled will be called archetype.

Note that according to this definition, which will be adopted here, the archetype is always abstract or immaterial.

2.2.3   Interrelations Among Important Terms within the Product's Lifecycle

Now let us return to the modelling of real objects. Possible modellees (*i.e.* the objects to be modelled) originate either in nature or from organized manufacturing. An interesting question in the latter case (no matter whether the products are under development or are already mature) is which one is primary, the product or the model? This question is similar to the famous question about the primacy of the egg and the hen. Different viewpoints can obviously lead to different answers, but what remains viewpoint-independent is that there is a relation or connection between the two. To avoid contradictions and reduce the uncertainties related to these terms, let us consider Figure 2.18.

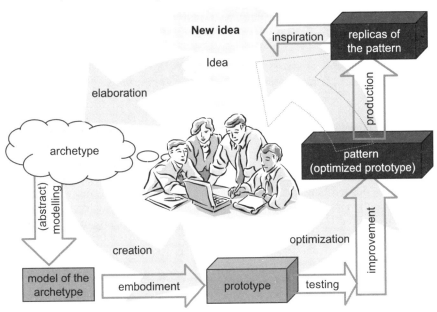

Figure 2.18. Interrelations among idea, model, archetype and prototype during the development cycle of an original product

Normally, at the beginning there is an idea. It is elaborated until there is enough information in it to prepare an archetype on its basis. On the basis of the archetype, a model is prepared. After that, through some manufacturing process, the model is embodied into a prototype. The prototype, in turn, is tested, improved and optimized until it can be used as a *pattern* for mass production. During the production, many *replicas* of the pattern are made. Any of these replicas, their production itself, as well as the model, the prototype or the pattern could inspire new ideas or be used directly as modellee and thus initiate the lifecycle (or the development cycle) of a *derived product*, as illustrated in Figure 2.19.

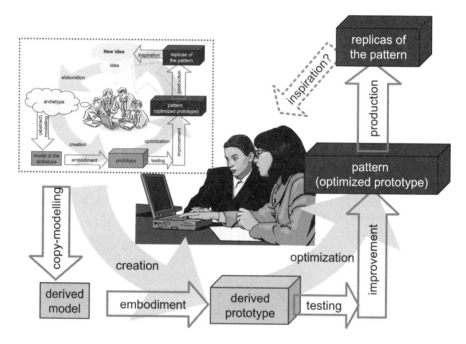

Figure 2.19. Interrelations between idea, model and prototype during the development cycle of a copied or derived product

In certain cases, derived developments are desired and intended, in other cases they are not desired but hardly avoidable, and in the worst case copy modelling or copy production could even be illegal. To summarize, copy modelling is not necessarily a breach of copyright or stealing of intellectual property – it actually takes place in each development loop and helps to improve the initial prototype. As "normal" modelling – shown in Figure 2.18 – has a lot in common with inventing, it can also be named *inventive modelling*.

Consequently, the answer to the question at the beginning of the section is that for inventive modelling the idea is prime, while for derived modelling the product (or object) is prime. This is true especially at the beginning of both processes – more precisely, for their first loop, since already for the second loop of the cycle, the case could become either mixed or transform into the "opposite" kind of modelling.

2.2.4   Process Models

Nothing endures but change

Diogenes Laertius
Lives of the Philosophers

2.2.4.1 Ambiguity of the Word "Process"
Similarly to the terms "model" and "modelling", the term "process" is used in many different contexts and refers to activities in totally different domains. In this text the focus will be on production-related processes (including modelling itself). We shall understand *product* to be a really existing object that is usually a result of a manufacturing process. We shall understand a *product model* to be a model of an existing as well as not-yet-existing product. Although virtual objects like software models, software, *etc.*, can also be the result of "manufacturing" processes, they will be referred to with their specific names.

Definition 2.8:   A process *is any non-empty and time-dependent sequence of interactions of two or more objects, leading to changes in (the state of) at least one of the objects.*

According to this definition processes are, for instance, the movement and changes in the orientation of an object (they can happen only as result of an interaction with the surrounding objects and are relative to them), changes in the structure or in the form of an object (usually as a result of applying mechanical, chemical or other forces). In the context of this work, we normally understand process to be a production-related one.

There are three terms in Definition 2.8 that could need more discussion: *event*, *time* and *interaction*. We shall define the term *event* as any ascertainable change in the state of an object or system of objects. The second term, *time*, has a tight relationship with the events and helps us distinguish one event from another. The third term is *interaction*, usually defined in terms of an action causing a reaction: in particular, as a pair comprising the action of one object on another one, and the respective reaction (or answer-action) of the affected object. It could be useful to distinguish here two different meanings, though. In the context of Newtonian physics, the action and reaction are two forces that (can) exist only simultaneously; in the context of process control, it sometimes makes sense to view the action and the reaction as two events that are related, but – in general – happening at different times.

2.2.4.2 Time

Time is an illusion. Lunch time doubly so.

Douglas Adams
The Hitchhiker's Guide to the Galaxy. Chapter 2

The main attribute of all processes, common to all of them, is their dependence on time. Although many great scholars have written about time, we shall try to introduce a short and pragmatic definition of this term:

Definition 2.9:   *For engineering purposes* time *can be viewed as an invisible but inherent characteristic of the universe, allowing us to correlate different events or processes, order them in a sequence and quantify the "distance" between them.*

On the one hand, we try to accomplish each (production) process in the shortest possible time. On the other hand, even if it were possible to work infinitely fast, and thus reduce the accomplishment time to zero, this would be rather impractical, since without time (time=0) we would lose the order of events, ending thus in chaos. Therefore, when there is no need to consider relativity theory, we assume axiomatically that:

3. time depends on nothing (that we could influence);

4. time advances with constant, greater than zero speed;

5. time advances only forward;

6. there are no interruptions in time;

7. all processes are time-dependent.

When all these assumptions hold, we can describe process flows or sequences as functions of time.

2.2.4.3 Relations Between Product, Production and Process

On the one hand, every product is a specific type of object. On the other hand, products are output of some kind of production. Production itself is a process. Therefore, process modelling implies also modelling of objects, or in other words, modelling of production processes implies product modelling, too. A simplified model of a production process is represented in Figure 2.20.

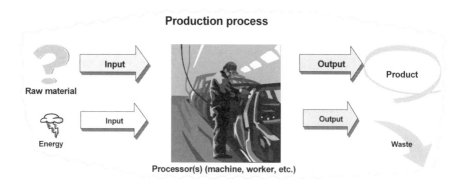

Figure 2.20. A simplified model of a production process

With regard to the relation between process and product, this simple model can be viewed from at least three different perspectives, depending on the starting point:

8. given the process, analyse what can be produced and what raw material is needed;

9. given the (required) product, find how to produce it and what raw material is needed;

10. given a raw material, analyse how it can be processed and what can be produced.

When two of the three elements are given, it is easier either to determine the third element or to make sure that no adequate third element exists.

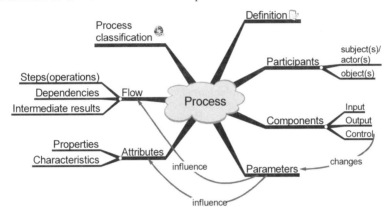

Figure 2.21. A model of a generic process

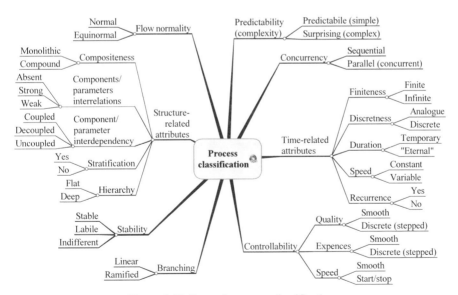

Figure 2.22. Example process classification

In contrast to Figure 2.20, which is normally "read" from left to right, the generic model in Figure 2.21 is focused in the centre, *i.e.* it is to be "read" towards the periphery, with the reading order – clockwise or counter-clockwise – playing no significant role.

Compared to a function in the mathematical sense, which is a mapping between two sets – input and output – the process represents a similar mapping between (the elements of) an input set and an output set, but it contains a time component, *i.e.* any process needs or takes time.

Processes can be classified according to different criteria. A very systematic classification from cognitive point of view can be found in Sowa (2000). Also, Sowa gives a very intuitive and mnemonic symbolic notation for different kinds of processes and for some of their elements – start, stop, branching, concurrency, *etc.* Another possibility, better suited (in my view) to the field of engineering, is illustrated in Figure 2.22.

The development of every product and every process is influenced by a few major factors. When they are restrictive, we can speak about limitations or constraints, Dörner (1987) even speaks about barriers. An example of influencing factors is represented (in a self-explanatory way) in Figure 2.23.

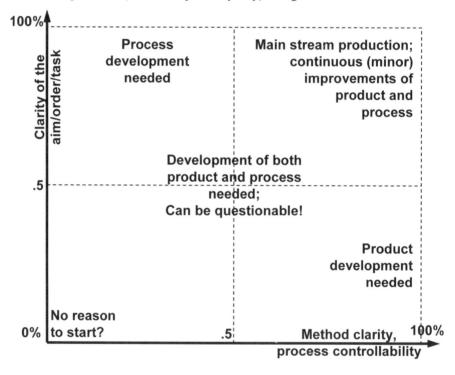

Figure 2.23. Role of some factors, influencing product and process development

In general, three main groups of questions should be asked in respect of the development of a (potential) product. They are given in Figure 2.24.

When trying to detail the answers to these questions it is usually helpful to prepare and consider a (meta-)model of the product development to consider the factors influencing it. An example is presented in Figure 2.25.

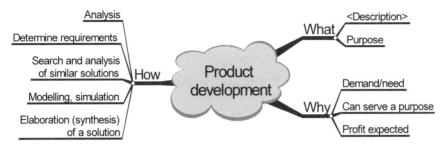

Figure 2.24. The main questions, related to product development and some related notions

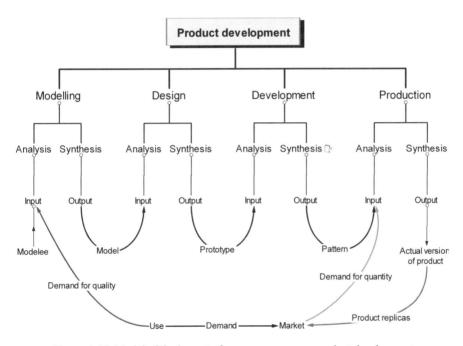

Figure 2.25. Model of the impact of some processes on product development

2.2.4.4 Simulation and Modelling of Processes

An explanation of the term *simulation* in a general sense is found in Wikipedia Wiki (2006) as *"an imitation of some real device or state of affairs. Simulation attempts to represent certain features of the behaviour of a physical or abstract system by the behaviour of another system."* A more complete definition of this notion that would be closer to our needs and understanding is adopted here from VDI-Richtlinien (2000)[6]:

---

[6] In original: "Unter dem Begriff Simulation versteht man ... die Nachbildung eines dynamischen Systems in einem Modell, um zu Erkenntnisse zu gelangen, die auf die Wirklichkeit übertragbar sind." (*cf.* VDI-Richtlinien (2000)).

*Definition 2.10: The implementation of a dynamic system in a model suitable for experiments **and the experimenting with this model**[7] to gain knowledge that can be transferred (back) to the reality.*

In this definition of simulation, we can distinguish the following essential points. The aim is not only to gain knowledge, but also to use it to act against weak points of the reality and towards its improvement. There are two phases in achieving this: (i) modelling of the dynamic system, and (ii) experimenting with the model. Consequently, the simulation relies upon modelling, where the modellee is a dynamic system.

2.3   Modelling: Classification

Known modelling approaches can be classified according to different criteria (*cf.* Figure 2.2), so let us consider some representative examples.

According to the medium used for the model, we can distinguish between *real* (mock-up) and *virtual* (mathematical, informational, software/computer models, *etc.*) models.

According to the application domain, we can distinguish between medical, psychological, linguistic, engineering, architectural, chemical, physical and other models.

According to the application (sub-)domain, we can distinguish (*e.g.*, within the engineering domain) between modelling in design, manufacturing, assembly, planning, marketing, service and others.

According to the basic modelling tool, there could be (CAx-) system-based, language-based (UML, Express, natural language, *etc.*), and other modelling.

According to the used (software) architecture, it is possible to have client–server architecture, distributed architecture, high level architecture (HLA) and so on.

According to the dominant method or approach, the modelling can be functional, object-oriented, feature-based, distributed, *etc.*

According to the involved concepts, the modelling can be modular, agent-based, holonic or other type.

It is also possible to classify the modelling according to the characteristics of the resulting model, which are to be guaranteed or expected.

A classification of several possible model types according to some key criteria is illustrated in Figure 2.26.

2.4   Model Traits

Ideally, each model would have or at least represent all the important traits of the modellee. In reality, the set of traits of the modellee and the set of traits of the model have a common subset, but are rarely identical (*cf.* Section 2.4.1.18 below). The traits that are specific to the model only, but not the modellee, can be called *model-specific traits*; they represent directly or indirectly the quality of the model. They depend on the modelling approach, on the methods used, on the chosen representation and many other factors.

---

[7]   The bold text is added from the author to make the idea clearer.

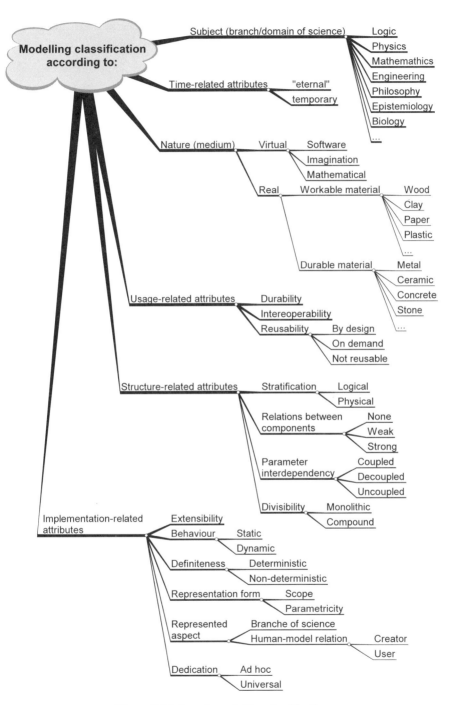

Figure 2.26. Sample modelling classification

The next section discusses some important model-specific traits that can be observed in most information models.

2.4.1   Definitions

It makes sense to define traits and their measurement in a way that will allow us to compare models of different types. This can be achieved if we use either relative values or a common comparison basis. For this reason, the formulae for calculation of the trait values have to be defined so that the range of the respective functions is between zero and one.

2.4.1.1 Compositeness

Theoretically, it is possible to distinguish between *atomic* (or *elementary*) models – that contain no other models – and *compound models* – that contain other (atomic or compound) models. Besides, "contain other models" refers not to the physical aspect but to the organization instead, especially in the case of software models.

It makes sense to define compositeness as having a Boolean value: *zero* for atomic modes and *one* otherwise. In practice, the software implementation of an atomic information model is not necessarily also atomic: for instance a point is – both as object in space and as (informational) notion – something single and undividable, but is usually modelled by means of at least two numbers – its coordinates in the coordinate system used. Similarly, the majority of software models are compound, too. The only exceptions are the models of some scalar attributes and properties of the modellees (like the point's coordinates above) – they are modelled by representing their value in a variable of an appropriate type (*cf.* Section 2.4.2.3.1).

The dependency of any trait listed below on the model's compositeness is to be explicitly mentioned in its definition.

2.4.1.2 Divisibility

This trait describes the possibility to split a given model in several sub-models that would be equivalent (as a group) to the initial model. Such activity can be helpful when optimizing composite (and complex) models to factor out a sub-model that is common to two or more models. In this sense the value is Boolean, it can only be true (*one*) or false (*zero*).

2.4.1.3 Accuracy

The *accuracy* is a trait, showing how similar to the modellee the model is, and what deviations can be expected. Extremely rarely, the accuracy of a model can be measured directly or calculated – typically, for very simple models only or for separate model traits representable with numbers. Since (the elements of) software models are represented by numbers, it is important to know how accurately the numbers can be represented in a computer. The accuracy of a software model depends on the hardware and on the representation used.

2.4.1.4 Actuality

The *actuality* can be viewed as a time-dependent accuracy, meaning that if a modellee is changing with time, its model should be updated or actualized in order to remain useful and fulfil its purpose. In a dynamic environment it is extremely important to work with the most recent information in order to be able to make proper decisions. On the other side, there is a desire to make the model's lifetime as long as possible (discussed in more detail later on – *cf.* Section 3.1.1.8). These two requirements seem contradictory at first glance – *i.e.* actuality of a model despite its long lifetime. Nevertheless, it is possible to achieve both of them relatively easily by well-directed and well-localized update of the non-actual components of the model. The actuality may change in two situations. On the one hand, it can "expire" if the modellee is changed or new information about it becomes available. On the other hand, new scientific discoveries can also make a model outdated and require its actualization. It can be necessary to assess the actuality in two cases: either to determine whether a given model needs to be updated or to determine which of two (non-numbered) versions of the model is more recent.

2.4.1.5 Adequacy

Whereas the accuracy reflects the mathematical and the numerical representation of a model, the *adequacy* reflects its logical and semantic correctness. Unfortunately, there is no way to measure or calculate this trait objectively.

2.4.1.6 Aspect

Depending on the purpose of the task at hand, each modellee can be viewed from different viewpoints or aspects. If a real three-dimensional object is viewed from two different viewpoints, some of the observed things can be the same, but most of them will be different or look different. Similarly, when modelling complex modellees, it can be useful to distinguish the inherent traits of their different aspects. For instance, when we are modelling the *manufacturability* of a given aggregate, we need to know all dimensions, the shape, the material, the pursued surface quality, among others. When another aspect is modelled, *e.g.*, the *functionality*, traits of the modellee like structure, connections among the components and others become more prominent than material and surface quality. Each different aspect of a given model is usually representable as a distinct layer (*cf.* Stratification).

2.4.1.7 Autonomy

The ability of a model, object or system to react to events and changes in conditions or environment in an adequate and purposeful way will be called *autonomy*. The autonomy is a way of self-control. It can vary on a scale from zero (fully controlled or fully dependent) to one – fully autonomous. This property is rarely inherent to atomic (informational) models. It is natural to expect autonomy of a computer model when the modellee is also autonomous. All four combinations of such a pair (modellee and model) with regard to autonomy are indeed possible.

The term autonomy refers mainly to the lifetime of the respective object (compare with independence below!).

2.4.1.8 Cardinality

In set theory, the number of elements in a given set is called *cardinality*. Analogically, this term will be used here to denote the number of sub-models or components within a model. Only the direct sub-models (*i.e.* without the nested sub-models) have to be counted[8]. We shall consider elementary models as having cardinality zero.

2.4.1.9 Changeability

This property is a measure inversely proportional to the effort needed to change the model. It makes sense to use absolute and relative changeability.

2.4.1.10 Compatibility

One entity (part, product, model, *etc.*) is said to be compatible with another entity (usually of the same type) if the former has functionality and properties corresponding to some degree with those of the latter entity. Apparently compatibility can vary on a scale from zero (fully incompatible) to one (fully compatible or equivalent), although a compatibility below 50% should be classified, in general, as "incompatibility". Some aspects of compatibility are given in Figure 2.27. Depending on purpose and application domain, though, some of these aspects become more prominent, others become negligible, but almost always one turns out to have a dominant importance.

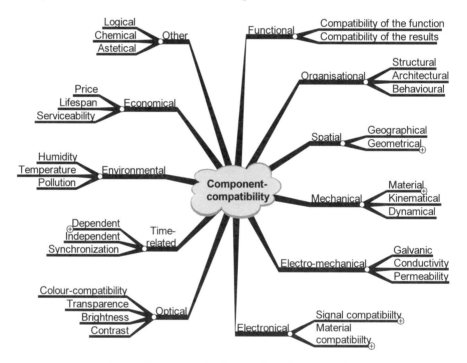

Figure 2.27. Example classification of compatibility

---

[8] In contrast to set theory, though, some cases exist where the cardinality of complex models should be calculated as the sum of the cardinalities of all components, applying this rule recursively, if necessary.

2.4.1.11 Consistency
When all components (or sub-models) of a compound model are described and represented without contradictions – *i.e.* they follow the same approach, use compatible methods and organization – we shall say that the compound model is *consistent*.

2.4.1.12 Dimensions
Computer models have only one dimension – their size, measured in bytes or derivative units. Physical models (mock-ups) can have many dimensions. Nevertheless, when a model represents (or "derives") some dimensions of the modellee, we speak about "x-dimensional-model" (*xD*-model), where *x* is usually a number from 2 to 4.

2.4.1.13 Durability
Durability is the ability of something to last. For materials, it depends on the material properties and on the conditions of use (environment, *etc.*). For immaterial things like concepts, ideas and similar, it depends on the durability of the respective host (see below for definition), medium or representation. Although we typically strive for high durability, it is not always good, since it could impair other traits like changeability, extensibility, flexibility and updateability (see below for their definitions).

2.4.1.14 Dynamics
The possibility for a model to (frequently) change appearance or behaviour is called *dynamics*. It is apparent that the model of any process will be dynamic. It is difficult to measure the dynamics. Although at the first glance it seems that there exist objects or models that do not change with time and thus should have dynamics=0 (or are static), a more careful look suggests that they actually contain two (or more) phases in their lifecycle that have different dynamics: *creation* (or genesis) – with dynamics=1 – and *post-creation* (or use, or existence) – with dynamics=0. Therefore, this trait is *time-span-dependent*.

2.4.1.15 Extensibility
This property describes whether it is possible to extend the respective object (model, product, *etc.*) with new functionality or other characteristics. Ideally it should be possible to infinitely extend anything (this could be denoted as extensibility=100%), but in the worst case extensibility is impossible (extensibility=0%).

2.4.1.16 Flexibility
In IEEE (1991) *flexibility* is defined as *"the ease with which a system or component can be modified for use in applications or environments other than those for which it was specifically designed"*. For the case of modelling we should redefine flexibility as the *ease with which a model or a system of models can be adapted for (use in) purposes, not intended or foreseen during the initial development*. Note that a new purpose may require new application, new environment or both.

In some cases flexibility can be achieved only by means of extensions (*cf.* the definition of extendibility above). In cases when flexibility is inherent without need to implement extensions, the term *versatility* is used as a synonym.

The flexibility can be expressed with numbers between 0 (enormous effort for any adaptation) and 1 (no effort for adaptation to an infinite number of purposes). Yet, it can be neither directly measured nor easily calculated. Nevertheless, we have to distinguish the flexibility of a system or compound object (respectively – compound model) from the sum of the flexibilities of its components. The former is usually much lower than the latter! The point is that the purpose is a determining factor, but for a system is defined top-down, while the system's embodiment happens bottom-up. Thus, using a given system with a new purpose would usually mean having a new set of functional requirements, which means that many existing modules would remain unused.

High reusability does not mean automatically high flexibility – if the object in question is reused again and again for the same purpose, it is simply durable, but not yet flexible. If a system can be used for a new purpose without adaptation, this means that either the new set of requirements is a subset of the old requirements, or the system exhibits great lateral functionality (*cf.* the respective section below).

2.4.1.17 Functionality

The *functionality* describes all capabilities of the model. It can be viewed as a set of all functions that a given entity can accomplish. Thus, it can be represented by the cardinality of this set, which is a number between 0 for no functionality and infinity for infinite, endless functionality. Actually, zero functionality would mean that the respective object is of no use, or – when the object is a model – that the model cannot fulfil its purpose. Therefore, the zero remains excluded from the range. The other end of the range – the infinity – is excluded too, since no object or model can accomplish an infinite number of functions. Thus, even the highest functionality will be a huge but countable number.

The requirements for any artefact depend on its purpose and can also be described as a set of functions, which form the *required functionality* (F_{req}). Since not every artefact fulfils its requirements, the "normal" (or *full* or *actual*) functionality is sometimes called *implemented functionality* (F_{impl}). It intersects the required functionality, as illustrated by the Venn-diagram in Figure 2.28.

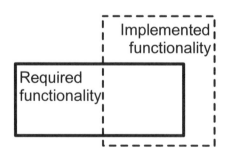

Figure 2.28. Functionality types

Functionality that is required, but not implemented, remains *due functionality* (F_{due}). It can be expressed as:

$$F_{due} = F_{req} - F_{impl} \qquad (2.1)$$

Functionality that is not required, but implemented – perhaps, due to specificities of the development process or due to other considerations – can be called *lateral*[9] *functionality* (F_{lat}). It can be expressed as:

$$F_{lat} = F_{impl} - F_{req} \tag{2.2}$$

Note that the subtraction in Formulae 2.1 and 2.2 operates on sets and is different from the normal subtraction.

The cardinalities of the respective subsets can be expressed as:

$$|F_{due}| = (0, |F_{req}|] \tag{2.3}$$

and

$$|F_{lat}| \in [0, |F_{impl}|] \tag{2.4}$$

Lateral functionality is very welcome when achieved as a side effect of the development (*i.e.*, without extra effort or costs), since for some new purpose it can become a required functionality and therefore increases the flexibility (*cf.* the definition in Section 2.4.1.16 above).

When the required functionality F_{req} is the same as the implemented functionality F_{impl} or the former is a subset of the latter ($F_{req} \subseteq F_{impl}$) it is said that the artefact fulfils (completely) its purpose. In such cases the cardinality of the subset of all due functions is zero ($|F_{due}| = 0$).

From the point of view of the importance of the specific functions that build the functionality, we can group them in two categories or subsets: *basic functions*, which implement the inherent functionality, and *auxiliary functions*, which implement functionality of lower importance.

When still unknown objects or artefacts are investigated, one can distinguish between *apparent functionality* and (yet) *hidden functionality*.

2.4.1.18 Coverage

Given a modellee and its model made for a certain purpose, the following considerations can take place:

11. Only the important for the respective purpose attributes and functions of the modellee have to be modelled (*cf.* Definition 2.1) and thus represented in the model.

12. Typically, some attributes and functions of the model (would) concern only the model itself and not the modellee.

13. Often there are attributes or functions that are not required, but nevertheless modelled.

If we try to represent the set of attributes and functions of a modellee $A_{modellee}$ and the set of attributes and functions of a model A_{model} as a Venn-diagram, the result is illustrated in Figure 2.29. Apparently, the greater the intersection of the two sets, the better (approximation of the modellee is) the model. We shall call the

---

[9] Other possible terms for this notion are *side* or *excess* or *extra* or *unrequested functionality*.

intersection *coverage*, since it reflects to what degree the model "covers" attributes and functions of the modellee, and shall measure it as percentage.

$$Coverage = \frac{\left|A_{\text{modellee}} \cap A_{\text{model}}\right|}{\left|A_{\text{modellee}}\right|} \tag{2.5}$$

Unfortunately, coverage gives a more quantitative than qualitative impression about the model, since the importance of single attributes and functions is different. Therefore, covering a large number of unimportant attributes and functions can be worse than covering a much smaller but more important number of them.

Similarly to the functionality, the coverage can be represented by Venn-diagrams as in Figure 2.29.

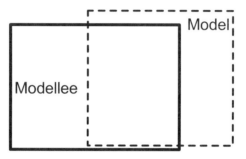

Figure 2.29. Coverage and suitability of a model

Clearly, better coverage means higher quality of the model. But with the increased cardinality of the set mentioned in 12 above, the *inefficiency* of the model also increases.

Now let us consider the coverage of compound models.

2.4.1.19 Compound Models
The traits of any model depend on the traits of its components. Unfortunately, very few model traits are representable as a sum or superposition of the respective traits of the components or by means of a simple formula.

Obviously, the set of all modelled attributes and functions can be expressed as

$$A_{\text{model}} = \bigcup_{i=1}^{M} A_{\text{sub-model}\,i} \tag{2.6}$$

Similarly, when the modellee is also compound, its respective set can be calculated as

$$A_{\text{modellee}} = \bigcup_{i=1}^{N} A_{\text{component}\,i} \tag{2.7}$$

Since in both cases overlapping between the sets of attributes and functions of the components and, respectively, of the sub-models can occur (*cf.* Figure 2.30), the cardinality of the top level sets will be smaller than or equal to the sum of the cardinalities of the components. The following formulae hold:

$$|A_{\text{model}}| \leq \sum_{i=1}^{M} |A_{\text{sub-model}_i}| \qquad (2.8)$$

and

$$|A_{\text{modellee}}| \leq \sum_{i=1}^{N} |A_{\text{component}_i}| \qquad (2.9)$$

finally,

$$Coverage_{max} = granule\_count * granule\_area \qquad (2.10)$$

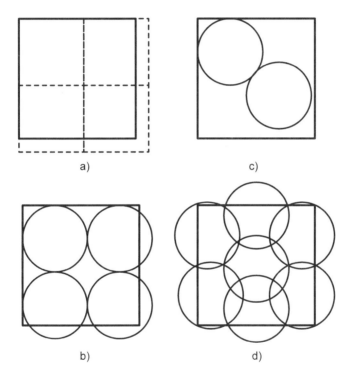

Figure 2.30. Coverage and granularity of compound models

2.4.1.20 Granularity (Only for Compound Models)

According to the Langenscheidt dictionary, the granularity is "*a measure for the size of the standalone sub-operations in which a process or program can be divided for achieving parallel processing*"[10].

---

[10] In the original: "**granularity** 1. Körnigkeit f; 2. Maß für die Größe selbstständiger Teiloperationen, in die ein Prozess oder Programm für die Parallelverarbeitung zerlegt werden kann". Langenscheidt Fachverlag GmbH, München, 1999

In the context of modelling, granularity refers to the average size of all sub-models of a given model. Since some models could have different dimensions, it is important to specify to which of them the word "size" refers in the previous sentence, *i.e.* which of them is taken for determining granularity. For instance, if the quality of the modelling is assessed, an appropriate measure for the "size" could be the coverage of each sub-model. If the efficiency of the memory usage is assessed, a better candidate for the dimension to be used could be the size of each sub-model in bytes.

Since the granularity can be very useful for comparison or assessment of compound models, it will be discussed again later on.

2.4.1.21 Homogeneity

This property shows whether all sub-models have the same (type) of origin and are thus homogeneous and directly compatible with one another, or have different (types) of origin and are heterogeneous. Sub-models of the latter type typically require special effort for their integration.

2.4.1.22 Independence

This is a measure of the strength of the relations to or of the dependencies on other elements of the surrounding system or environment. It could be related to or combined with model properties like existence, functionality and others.

Unlike autonomy (*cf.* the respective section above), independence is more related to the genesis of an object than to its lifetime.

2.4.1.23 Intelligence

This property is discussed in Section 2.4.2.3.1.

2.4.1.24 Interchangeability

If two entities (real or virtual) are fully compatible with each other (*i.e.* equivalent) and each can be used instead of the other without discernable loss of functionality, quality or anything else, we say that they are interchangeable. When a single entity is said to be interchangeable, it is meant that the design of the entity provides such a possibility and that spare parts of the same type are deliverable. Interchangeability is usually viewed as a binary (*i.e.* true or false) property (*cf.* also *compatibility* above).

2.4.1.25 Openness and Modifiability

The term *openness* refers to the possibilities of changing or extending any given model, and is *implementation-dependent*. The less functional a given model is, the higher is the probability that new desires concerning its functionality will arise, so that the model will have to be extended. The more complex a given model is, the higher is the probability that errors will occur or (for mechatronic systems) failures will happen during the exploitation, so that the model will have to be corrected/changed/repaired at the end user's place. For pure software models this is seldom a problem, but for complex mechatronic systems the distance to the place of use could cause problems (or at least additional costs).

Increasing the openness of a given model has strong influence on many of its other traits. In most cases it is positive – extendibility, flexibility, integrability, *etc.* In one aspect, though, the change is negative: the increased openness of a model makes it easier for the competition to imitate. For this reason many producers of software and software models sell their products as turnkey products. A necessary condition for achieving model openness is a clear definition of its interfaces. Depending on the model type, the interfaces can be mechanical, electrical, software, or any combination of these.

2.4.1.26 Paradigm

Webster's dictionary gives the following definitions for paradigm:

> *Example, pattern; especially: an outstandingly clear or typical example or archetype.*
>
> *...*
>
> *A philosophical and theoretical framework of a scientific school or discipline within which theories, laws, and generalizations and the experiments performed in support of them are formulated.*

Another explanation is found in Wikipedia: *"From the late 1800s the word paradigm refers to a thought pattern in any scientific disciplines or other epistemological context."*

One of the most popular paradigms in modelling is the *object-oriented modelling* (or *object-oriented paradigm*), related also to *object-oriented analysis*, *object-oriented design* and *object-oriented programming*. The main objection to all OO techniques is that the attribute "object-oriented" is somewhat misleading. Actually, the focus of these techniques is the grouping of similar objects into classes in order to factor out the common knowledge (data, procedures, *etc.*) about them and to increase the efficiency through reuse. Therefore, an attribute like *class-oriented* would be more self-explanatory.

2.4.1.27 Platform

By *platform* we shall understand the set of hardware, operating system and possibly additional software, providing an environment for a software product or software model to "live in".

2.4.1.28 Portability

Portability is usually defined as the easiness of making a model usable on a different platform, *cf.* for instance Howe (2006). For the purposes of computer aided engineering I would define it as the (average) easiness of making a model usable on any possible platform. It is inversely proportional to the effort necessary to adapt the model for use on a new platform. This effort is proportional to the number of platforms and to the complexity of the model to be adapted. So it is not trivial to compare the portability of differently complex models.

One of the ways to achieve (better) portability is to develop the models upon a layer (or basis) that is already portable.

2.4.1.29 Effort for Porting to new Platform

Very often it is necessary to make already existing model available and functioning on a new platform. The process is called *porting* or *migration* and the additional work to achieve this is the *effort for porting* (*cf.* Figure 3.5 in the next chapter).

2.4.1.30 Platform Independence

We shall understand by platform independence, the ability of a software application or software model to run on different platforms without (or with minimum) changes for adaptation.

According to Frankel (2001, p.25), the notions of main importance to achieve platform-independence have evolved from the 1960s until now starting from processors in the 1960s, through 3GLs (third generation languages) in the 1970s and 1980s, through middleware in the 1990s and MDA (model driven architecture) since the new millennium.

2.4.1.31 Quality

It is difficult to measure quality since it depends on the purpose of the model. The same model can be perfect for one purpose but totally unsuitable for other. For this reason, no mathematical definition will be given, but let us consider one guideline of The Association of German Engineers (cf. VDI-Richtlinien (1993)):

> *The quality of the model is decisive for the quality of the analysis results. Only if the model realistically describes the system, it is possible for the subsequent model analysis to produce results that can be transferred to reality.*

2.4.1.32 Reliability

According to Howe (2006) reliability (of a system) is "*An attribute of any system that consistently produces the same results, preferably meeting or exceeding its specifications. The term may be qualified, e.g. software reliability, reliable communication.*".

$$Reliability = F\left(reliability_{Host\ system}, \prod_{i=1}^{N} reliability_{component\ i} \right) \qquad (2.11)$$

2.4.1.33 Reparability

During its use any model can get broken, malfunction or cease to be useful. This could happen due to internal problems (specific mainly to material models) or due to a critical change in the environment. The latter can impact on all kinds of models, including software models – a typical example was the problem of the 2000th year[11]. We shall call *reparability* the possibility to restore the functionality of the respective model or the conditions allowing us to use it in accordance with the initially foreseen purpose.

2.4.1.34 Reusability

Before defining the term "reuse" as "second or multiple use of something", let us see what we mean by "(first) use". Some reasons for ending the "first use" of a product are listed below.

---

[11] This problem caused calendar-related modules in some improperly designed software programs and electronic devices to function improperly due to overflow of the year-counter.

a) There is no more need to use it for the initially foreseen or intended purpose.
b) It gets broken or there is a malfunction.
c) There is no qualified user anymore.
d) Its use is not legal anymore.

In case b) we can say that a *purposeful* or *natural reuse* can be pursued. In case a) one could try to reuse the product for alternative purposes. Since the aim is to achieve economic advantages or even profit, a suitable term here can be *economically based reuse*.

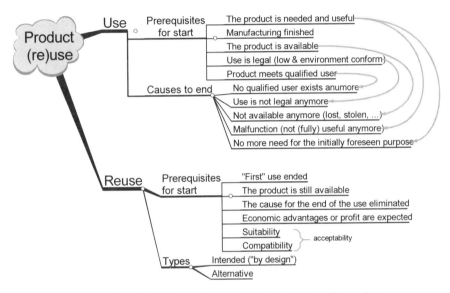

Figure 2.31. Conditions and prerequisites for (re)use of a product

Depending on the flexibility of the product and – with a composite product – on the flexibility of the components, several possibilities have to be considered. Of course, in the best case every product is fully reusable and in the worst case – absolutely not reusable. In the majority of cases some subset of the components of any composite product can be reused.

$$Reusability = f\left(\sum_{i=1}^{N} \left(R_{Component_i} \right), \sum_{i=1}^{N} \left(\sum_{j=i+1}^{N} \left(R_{Link_{i,j}} \right) \right) \right) \qquad (2.12)$$

2.4.1.35 Robustness

In general, this term is used to denote the capability of a product to keep its integrity, usability and functionality despite negative and possibly unforeseen influences of the environment. In the context of computer science, the meaning is extended to cover not only hardware but also the software, including behaving incorrectly or – possibly intentionally – even illegally.

2.4.1.36 Scalability
When a greater need for a given function can be satisfied by employing more instances of the respective model (or object), working in parallel, we shall speak about *scalability*. This trait belongs more to the result of a process than to the respective "processor" and does not always exist. For instance, we can use more cars for transporting more things, but more cars cannot help us travel faster – we need another quality. In other words, we could define the scalability as the possibility to trade quantity for quality.

2.4.1.37 Size
The size of a software model is the volume of memory it needs to be saved on disk – sometimes referred to as *static size* – or in the operating memory – also known as *dynamic size*. It is measured in bytes or their derivatives. The size of a software model is also an indirect measure of its complexity (*cf.* Section 3.1.2 in the next chapter).

2.4.1.38 Time Dependency
Time dependency explains whether a given model changes with time or not. It makes sense to define this trait as having a binary (or Boolean) value, since atomic models are either time-dependent or not. A compound model becomes time-dependent if any of its elements is time-dependent. All models of processes are time-dependent, too.

2.4.1.39 Universality
This property shows for how many different purposes a given model is well suited. More purposes mean higher universality and *vice versa*. Although it can be tempting to develop models with universality as high as possible, it is not always rational to do so. And achieving a full universality – *i.e.* developing a model that is suitable for all thinkable purposes – is impossible.

2.4.1.40 Updateability
The more valuable the trait actuality of a given model is, the more important becomes the possibility to update this model regularly. The actuality of a non-updateable model (*e.g.*, a wood mock-up) can only decrease with time, while the actuality of an updateable model can be improved regularly and on demand.

2.4.2   Organization of Models

In order to understand models, their capabilities and interaction it is important to analyse how they are organized. The two most often used terms in this respect are *structure* and *architecture* of models. Some of their more important properties, as well as their interrelations, are visualized in Figure 2.32.

Static models (*e.g.*, a physical mock-up of an object) do not have organization – they have only structure and sometimes also architecture. Monolithic objects have neither structure nor architecture, but they still can have static functionality and relations with the environment. Software models are usually dynamic models, having always structure, functionality, architecture, external relations (interfaces), and are unique in the sense that they have a programmable behaviour. For these reasons we shall say that they have organization too. Many authors use the terms organization and architecture as synonyms, but for the above-mentioned reasons I consider the architecture a component of the organization.

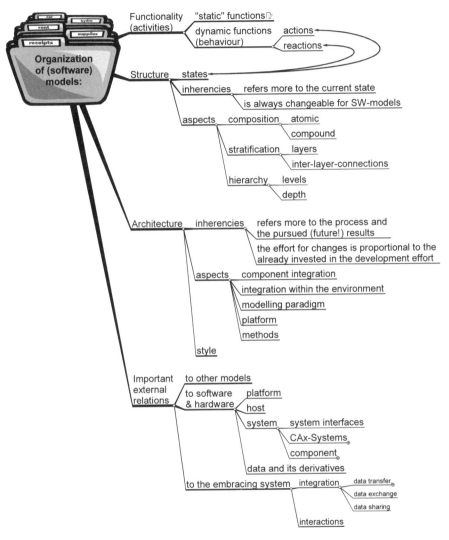

Figure 2.32. Organization of software models

2.4.2.1 Structure
The structure of a model describes its components and the (static) relations among them. Only compound models have structure.

2.4.2.1.1 Inherencies
The structure of a model refers mainly to its current (or already achieved) state in the model's lifecycle. It can be (and usually is) changed between the iterations of

the development process. The structure of a software model is always changeable, but this is not true for existing non-software models, *e.g.*, for mock-ups.

2.4.2.1.2 Structure-related Aspects of Models

The structure can be viewed from several viewpoints or aspects.

2.4.2.1.2.1 Composition

This attribute describes how the models are built. The simplest models are the so-called *atomic* or *elementary models*. They are *monolithic* or *not dividable*, having thus no structure but being used in building all other (non-atomic) models. Atomic models are often used for modelling properties of a modellee that are representable as scalar values.

Assuming that several models are available, they can be *grouped* according to different criteria. A group can be useful for easier reference to all objects simultaneously or for defining and performing operations on all elements of the group at the same time. Depending on the purpose, different criteria can be chosen. Grouped models can come from the same library or developer. They may be independent from one another or even be hosted on different platforms.

Models comprising other models are said to be *compound models*. For instance, the model of a circle in Figure 2.17 has (among others) the properties radius and centre point. The former property can be viewed as a model of a scalar, the latter as a model of a point; therefore the model of the circle itself is a compound model. Any compound model can be viewed as a group of models (*e.g.*, the radius and the centre point can be viewed as a group of models, belonging to the same compound model), but not every group is a compound model – *e.g.*, the group of all models, created by the same modeller are not necessarily parts of a (larger) compound model.

Often we have to model several objects (or modellees), which do not belong to one another but are interacting within a given process or are somehow related, and we need to model this relation. In such cases, we speak about *system of models*. If the involved models are physically or geographically distributed, but still interact with one another, they can be referred to as a *distributed system of models* or a *system of distributed models*.

2.4.2.1.2.2 Stratification

Another structure-related aspect of models is their stratification. It is inherent to some compound models and reveals the existence of several *layers* in their structure. The criteria for stratification are very interesting from a scientific point of view, *cf.* Figure 2.33.

Since every model is similar to the modellee, it is natural for a model to be layered if the modellee is layered too. Nevertheless, all combinations between the two are possible, as illustrated in Table 2.3.

In some cases the modellee can be viewed from different viewpoints and thus exhibit different *aspects*. From an organizational point of view the aspects can be represented either as different layers in the same model or as different standalone models. An important characteristic of the different layers is that they are always connected in a certain way. In the example for case #2 in Table 2.3, for instance, the values of the three colour components are connected with one another so that they form together the modelled colour.

In general, there are properties that are represented in a single layer, in all layers, or in a subset of layers. We shall call the modellee properties that are present in (almost) all layers *core properties*, those that are present in at least two layers – *essential properties*, and the rest – *aspect-specific* or *auxiliary* properties. It is obvious that the core and the essential properties can be used to build *inter-layer-connections*. They play an important role in integration of separately modelled aspects.

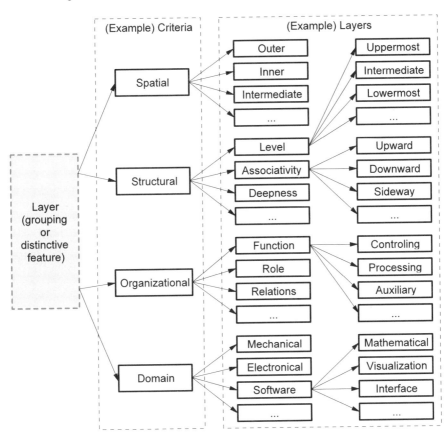

Figure 2.33. Criteria for defining layers

2.4.2.1.2.3 Hierarchy

The hierarchy is an attribute of the structure, which describes the systematic character in the order of the components and their relations. It can be said that a hierarchy is vertically divided in several *levels*, whereas each level can contain one or more components. In many cases we have to do with nested structures of models, in which hierarchical structures are clearly recognizable. A simple hierarchy is given in Figure 2.17.

Table 2.3. Stratification of modellee and model: possible combinations

#	Stratification of the:		Possibility	Plausibility	Example
	modellee	model			
1	no	no	yes	yes	sculpture of a man
2	no	yes	yes	yes	all colour models (RGB[12],CMYK[13])
3	yes	no	yes	yes	clay model of a car
4	yes	yes	yes	yes	car and a model car

2.4.2.1.2.3.1 Levels of Hierarchical Structure
Levels are the biggest components of each hierarchical structure. Splitting of sophisticated systems, models or objects in several levels, where each level consists of approximately the same number of objects to deal with, is an often used method for hierarchical structuring. It simplifies the manipulation of both the separate levels and the objects within each level.

The same holds for sophisticated models, but sometimes the number of levels in the model is not equal the number of levels in the modellee – it can be both greater or smaller. One example of a "Multi-dimensional Meta-modelling Architecture" is given in Jeckle (1999, p.11). He describes five level of a modelling hierarchy:

- M^{-1}:   Instance
- M^0 "Reality"
- M^1:   Modelling language
- M^2:   Meta-language ("Grammar")
- M^3:   Meta-meta-language ("Meta-grammar")

Again there is discussed the use of the XML Metadata Interchange Format (XMI format) within the four layer Meta-model Architecture Jeckle (1999, p.14).

Depending on the purpose and the point of view, the same structure can sometimes be interpreted either as layered or as hierarchical (levelled). For instance, depending on distinctive characteristics, the group of models clothes(skin(muscles(skeleton))) can be viewed either as nested, hierarchical structure or as models belonging to the different layers beauty, protection, movement and support.

2.4.2.1.2.3.2 Depth of Hierarchical Organization
This characteristic can give us an impression of the complexity and the possible minimal and maximal number of elements in a hierarchy. It can be used either to show the relative position of an element or to denote the depth of the whole hierarchy. The model in Figure 2.17, for instance, exposes a hierarchy of depth 6.

---

[12] RGB: method for representing any colour as a mix of the primary colours red, green, blue.
[13] CMYK: A colour model that describes each colour in terms of the quantity of each secondary colour (cyan, magenta, yellow), and "key" (black) it contains. The CMYK system is used for printing Howe (2006).

The minimal possible depth is one, which implies that the organization is *flat* or there is no hierarchy. There is no restriction on the maximum depth, but the complexity of organization rises exponentially with each new level in the hierarchy.

2.4.2.2 Architecture
This term is so ancient that it is usually viewed as already known and is rarely defined. One of the few definitions that can be found is given in Wikipedia Wiki (2006):

> Architecture (in Greek αρχή = first and τέχνη = craftsmanship) is the art and science of designing buildings and structures.

The literal translation makes an allusion to the fact that many creators have imitated already existing artefacts, since this is easier than creating a totally new artefact. Often the first artefact has already established a pattern or even a norm that is simply followed by the others.

2.4.2.2.1 Inherencies
We can think of three properties that are inherent to any architecture. It defines how the subcomponents one level below the top level are to be put together. This depends on the pursued results. In fact, changes are always possible, although more and more difficult with time.

2.4.2.2.2 Architectural Aspects of Models
Several aspects of model architecture are clearly distinguishable: the model structure, the integration within the environment, the modelling paradigm, the platform and the methods to be used.

To summarize, the architecture can be defined as the combination of concepts, approaches, methods and techniques that are used in the initial building phase of any model, system, or other compound and complex object, and therefore plays a crucial role in determining its most important traits and its entire future – development, use, maintenance, re-use, *etc.*

2.4.2.3 Important External Relations

2.4.2.3.1 Relations (of Software Models) to Data and its Derivatives
On the lowest level of all kinds of information and knowledge structures is the data. Each of the levels above can be viewed as a data derivative. Since these derivatives play an enormous role in modelling, we shall explain briefly some of them.

2.4.2.3.1.1 Data and its Derivatives

2.4.2.3.1.1.1 Data
We shall understand data to be *strings of (ordered) symbols*. These symbols are represented in computers by numbers, and in turn, the numbers are represented by binary digits or bits. Apart from bits, the numbers are the smallest "building block" in the representation of data and data derivatives – including algorithms – in a computer. An example of such a string of symbols is "3.1415".

2.4.2.3.1.1.2 Meta-data
The meta-data is a connection or relation between groups or pieces of data. Given the strings of data "5.003" and "5.02", we can connect them into a relation, for instance by the sign "smaller than":

$$5.003 < 5.02$$

Thus, the "$<$" symbol has special meaning and is meta-data in this case.

2.4.2.3.1.1.3 Information
The information emerges from interpreting data. Interpreting means that each piece of data and meta-data is connected or put into a relation with already known facts as well as with all other pieces of data. Thus, the example above would be interpreted as putting two numbers in a relation, saying that the second one is greater than the first one. To achieve this, the interpreter (human or machine) should have some *a priori knowledge* – *i.e.* to be able to read the numbers and to understand the meaning of the sign "$<$". This knowledge is often called *context* or *background knowledge*.

Since the context always plays a crucial role by gaining information from given data, another possible definition for information is *data in certain context*. Thus, when hearing the (incomplete) expression "to be or not to be…" people, who are experts in different domains, can interpret it differently. A fan of Shakespeare could see an allusion to the famous phrase of Hamlet; a philosopher could think about The Question of Douglas Adams' book Life, the Universe and Everything, and a mathematician could write it down on a piece of paper as $2b \mid \neg(2b)$ and say "this is always true!".

Another possibility to define the term information is as a combination of meta-data and (groups of) data that are to be connected/related, for instance:

$$\pi \approx 3.1415$$

Such a combination of data and meta-data is usually named an *attribute-value pair*. In representing more complex information it is possible to nest attribute-value pairs by using a given pair as the value of another one. A sample graphical representation of nested attribute-value pairs can be seen in Figure 2.34.

When the value of an attribute-value pair contains just data (*i.e.*, there is no nesting), this pair can be named *basic* or *substantial* attribute-value pair. Independently of their representation, basic attribute-value pairs can be viewed as the smallest units of information.

2.4.2.3.1.1.4 Meta-information
Meta-information is information about other information. For instance, the node "Invariable properties" in Figure 2.34 is meta-information and denotes that all nodes below it are invariable for the whole class of objects of the circle type.

2.4.2.3.1.1.5 Knowledge
We shall call *knowledge* the ability to gain new information from already existing information or data. For instance, knowing that the circumference of all circles is equal to $2\pi R$, and that the radius of a given circle C is $R=1$, we can conclude that the circumference of C is equal to 2π.

Another possible way to define knowledge is to view it as the union of the meta-information and the pieces of information that have to be connected.

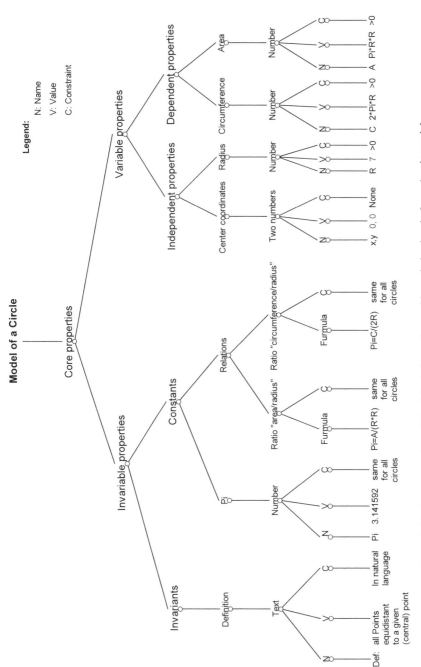

Figure 2.34. Representation of the data, information and knowledge levels for a simple model

2.4.2.3.1.1.6 Intelligence of a Software

Intelligence is a quality that is inherent mainly to human beings and is usually related to thinking and especially to the *ability for reasoning* on the basis of *a priori knowledge*.

We perceive and rate intelligence mainly based on behaviour. But what is, actually, behaviour? If we define the set of all activities of a given person or object as behaviour, it is possible to distinguish between two main behaviour types:

    c) *passive behaviour*: reactions to external influences (coming from the environment or from other objects);

    d) *proactive behaviour*: actions, initiated for some internal reasons and having – intended or not – impact on other objects (*i.e.*, causing their reactions);

Of course, a mix of both above-mentioned types is also possible and can be referred to as either *mixed behaviour* or simply *behaviour*.

Strictly speaking, it is almost impossible to meet pure proactivity as defined above. In practice, when we say about a human that he is proactive we usually mean that he shows initiative. In turn, taking the initiative from an individual can be viewed as attempt on his part to predict or anticipate the next action towards himself and prepare for it or even initiate the appropriate reaction in advance.

Since any reaction can cause another reaction, often one simple activity initiates a chain of interlinked actions and reactions – *i.e.*, interactions. From the point of view of an external observer, the life of any object can be viewed as a chain of interactions with its environment. Note that according to the definition above, activities having negligible or no influence on other objects (*e.g.*, breathing, digesting or even thinking itself), are not considered proactive. Both thinking and reasoning are themselves activities, internal for every individual. Therefore we can rate them and estimate how intelligent they are only after the result of the reasoning is communicated to us by some activity – at least by speaking.

Now let us turn to the software. On the one hand, it is obvious that any software possesses some (kind of) behaviour. On the other hand, the possessing/exposing behaviour alone is insufficient for being intelligent. In regard to humans we would say that somebody's behaviour is intelligent after we compare it with either another person's or with our own (supposed or real) behaviour in a similar situation. Thus, we can state that a) there is no absolute intelligence and b) *intelligence is relative* and can be discovered only in comparison with something. In regard to software, usually similar reasoning is applied: when we say that a program is intelligent, we mean that in a given situation it either behaves better than most programs with similar purpose, or attempts to behave like a human being who experiences a similar situation. The most well-known test in this area is the test proposed by Alan Turing and named after him Turing (1950). This test is based on a chain of questions and answers (interactions). It should help one to decide whether a given machine or program can think and, consequently, can be considered intelligent. Up to now, no computer/program has passed this test; why should we then discuss intelligence of a software systems? The point is that on the one hand, the Turing test gives only a binary answer whether a given computer (system) is intelligent; on the other hand, each travel begins with the first step and we have to do it, if we want to reach the destination. Therefore, we need some other measure of intelligence in the meantime until intelligent machines become available. For engineering purposes we do not have to start with fully intelligent machines.

Anything more than "not intelligent at all" can lead to improvements and savings in the respective area.

We shall view a product or process model as being intelligent if it possesses at least one of the following capabilities:

e) to behave/react as the modellee or simulate its reaction when the user simulates acting on it;

f) to guess what (re)action is desired/needed in any given moment and either propose alternatives or perform it immediately (cf. the description of initiative by human's behaviour above);

g) to find, determine or request any missing data or information alone;

h) to recognize invalid data or information and correct it alone.

The more of these capabilities a given model possesses, the more intelligent it is considered. On the other hand, of two models, the one that can complete more tasks with less effort of the controlling/requesting user or program is considered more intelligent. The effort could be measured either as number of necessary instructions or as the time spent to give them.

Apparently, neither models drawn on paper, nor models made of clay or other workable materials can be given behaviour and, therefore, they cannot become intelligent. The only model type to which intelligence can be granted is the software model.

In general, proactive behaviour would require more intelligence from a model than passive behaviour, since the initiative implies own desires resulting from thinking or reasoning. Computer science, though, is not expected to achieve such advances in the next decade. Therefore, we concentrate on passive behaviour, since it is simpler to implement. But as mentioned above, passive behaviour alone is not sufficient to achieve intelligence. Turing (1950) says:

> Intelligent behaviour presumably consists in a departure from the completely disciplined behaviour involved in computation, but a rather slight one, which does not give rise to random behaviour, or to pointless repetitive loops.

Two types of passive behaviour are distinguished: reactions to commands (leading to a desired result that is known in advance) and reactions to other events (all actions without commands). It is clear that reactions to commands ("disciplined behaviour"), which is immanent for every software product, cannot represent intelligence.

2.4.2.3.1.1.7 Wisdom

Wisdom is empirical information (experience), complementing the available knowledge and intelligence and making possible or increasing the probability to take proper decisions in yet unknown or not explicitly foreseen situations.

2.4.2.3.1.2 Models for Representation of Data and its Derivatives

We have defined data as strings of symbols, and each symbol is represented in the computer as an integer number. There are no (more) problems with representing symbols as numbers after the adoption of the Unicode standard, since the number of symbols used is finite. But how should the numbers themselves be represented when their number is infinite and no computer has infinite memory? Each pupil knows that the number of different fractions is infinite even in small ranges like this between 0 and 1. Then, how is the finite number of symbols, plus the infinite quantity of numbers, represented in the finite computer memory? The answer is simple: by analysing the different application areas, and using a model of the respective range of numbers, which is appropriate for the given purpose. These models are called *data types* and have the inherencies illustrated in Figure 2.35.

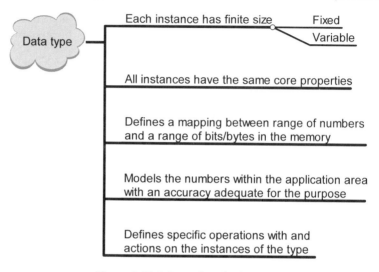

Figure 2.35. Inherencies of a data type

The simplest data type is dedicated to representation of Boolean values (true and false). Since there are only two possibilities, a single bit[14] would be sufficient, but due to technical considerations, usually a whole byte[15] is used. The representation of a finite set of consecutive numbers is commonly achieved by means of the data type *integer* or its derivatives. This data type has a typical size of 2, 4 or 8 bytes, depending on the implementation. The number N of consecutive numbers in the representable range R is directly dependent on the size S of the data type (in bytes) in the specific case, and is calculated with the following formula:

$$N = 2^{S*8}$$

(2.13)

Thus, with an integer data type variable of size one byte we could theoretically represent 256 consecutive numbers. If the numbers to be represented have a sign (*i.e.* the range is zero-symmetric), one bit has to be reserved for the sign representation. In this case, since the zero is "sign-neutral", *i.e.* can have both signs, the integer data type should either be able to represent +0 and –0, which is redundant. So one number less can be represented, or the implementation should perform some checks for representing +0 and –0 as the same state of bits in order

---

[14] A *bit* (abbreviated b) is the smallest information unit used in computing and information theory. It can be either one or zero and thus can represent any two mutually exclusive values or states like true or false, "on" or "off", *etc.*

[15] One byte (abbreviated B) has eight bits. Often used with prefixes like kilo, mega, giga, *etc.*

to avoid the reduction of the amount of representable numbers by one. The size of this data type is typically restricted to 8 (sometimes also to 10 or more) bytes, and thus the largest representable symmetrical range R is

$$R \in [-2^{8*8}, 2^{8*8}] \ or \ R \in [-2^{64}, 2^{64}] \tag{2.14}$$

or

$$R \in [-18446744073709551616, 18446744073709551616] \tag{2.15}$$

Although fairly large, this range could be insufficient for some applications. So, to represent either larger ranges or rational numbers with a finite number of bits, another data type is used: *floating-point number*. Assuming that we reserve one bit for the sign S, p bits for the significand[16], representing the most-significant digits of the number, and e bits for the exponent, the mathematical model of the range of numbers R representable by floating-point data type with *size*=$1+p+e$ bits will be as follows (all numbers in decimal base!):

$$R \in [-1^1 * 2^p * 2^{2^e}, -1^0 * 2^p * 2^{2^e}] \tag{2.16}$$

Here the significand is represented by 2^p, the factor 2^e is the exponent and the base is 2. The symbol $\in$ in the last expression should be interpreted with a "restriction": since p and e are integer numbers, not all numbers within the given interval are representable (*e.g.*, no number between 2^2 and 2^3 is representable and exponentiation is used in the representation of both the significand and the exponent), therefore R is only defined through the representable numbers. Thus, only a small subset of the real numbers can be represented in a (digital!) computer exactly, and the rest are represented as the nearest rational numbers. A more detailed description of the floating-point representation and its problems would go beyond the scope of this work, but can be found, *e.g.*, in Goldberg (1991).

2.4.2.3.1.3   Data-derivatives in Software Models
The software models are built-up from data, data-derivatives and (possibly) code. Therefore, many of the model traits depend on the traits of the underlying data and its derivatives, as well as on the chosen representation. For that reason, it should be kept in mind that most of the properties of software models are represented through real numbers, and when these numbers are approximated in their computer representation, the respective models could be badly influenced. Software models can use additional data for specialization (concretization) and communication. Software models can use bound or built-in code (as a special kind of data) for implementing intelligence.

2.4.2.3.2   Relations to other Terms and Software and Hardware Components
The information technology (IT) has huge influence on all computerized production methods. Since the rapid IT developments during the last decade have introduced numerous novelties and respective new terminology, let us consider some definitions and assumptions that would facilitate the further discussion.

---

[16] According to Goldberg (1991), "*This term was introduced by Forsythe and Moler [1967], and has generally replaced the older term mantissa.*"

2.4.2.3.2.1 Platform

The term *platform* denotes the hardware and software used as a basis for either development or use of a given software model. When a model is used on a platform that differs from the platform it has been developed on, we speak about *cross-platform development*.

2.4.2.3.2.2 Host

Since the platform for the use of one particular software model may differ from the platform for its development, the possibility to distinguish between these two platforms is crucial. The platform where a given model can be used (or where the model can "live") is referred to as *host*. The more platforms that can be used as hosts of a given model, the more portable or *platform-independent* this model is. The lower the number of components or layers required for the host, the higher the autonomy[17] of the respective model.

2.4.2.3.2.3 System

The term system is overloaded with different meanings related to different areas of the science. A definition of this term that resembles our understanding closely enough is given in Wikipedia (*cf.* Wiki (2006)): "*A system is an assemblage of inter-related elements comprising a unified whole. From the Latin and Greek, the term "system" meant to combine, to set up, to place together*". For our engineering purposes this definition has to be slightly modified in order to reflect the specifics of the majority of systems – both models and modellees – in the field of engineering.

Definition 2.11: A system is an assembly of inter-related components (subsystems, modules or elements) built together in a unified whole to serve a certain purpose.

In this sense, any compound model is also a system.

The purpose of a system together with the art of the components and their connections and relations determine most of the system's properties. A simplified model of a system, based on its most important properties, is presented in Figure 2.36. A taxonomy of some more important system-related attributes, terms and activities is presented in Figure 2.37.

2.4.2.3.2.4 System Interfaces

A system is typically connected to the outside world through *interfaces* – the set of all discernable input and output "channels" of the system that ensure cross-boundary communication with the outside world (*cf.* Figure 2.36). Apart from systems, all subsystems, modules and other components have interfaces, too. A system of hardware components has *hardware interfaces*, while a system of software components has, respectively, *software interfaces*.

Since the software "lives" in hardware (*cf.* Section 2.4.2.3.2.2 above), the properties of all software components – including interfaces – are strongly influenced by the properties of the underlying hardware. Software models either "live" in software systems or form themselves *systems of software models*.

---

[17] *Cf.* the definition in Section 2.4.1.7.

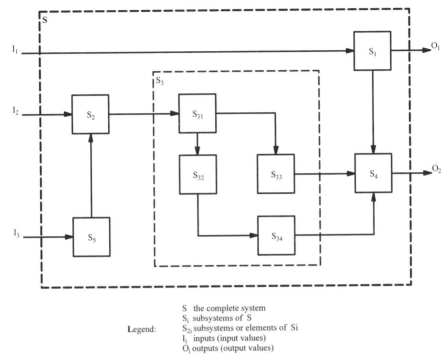

Figure 2.36. A simplified model of a system, advanced after Pahl and Beitz (1993)

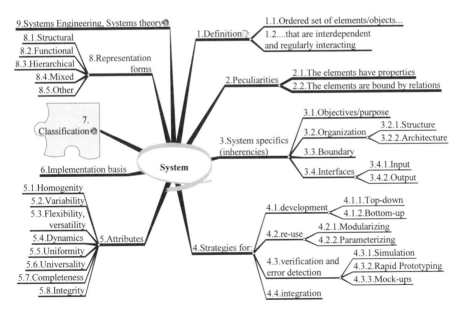

Figure 2.37. A taxonomy of some system-related attributes, terms and activities

In the domain of software programs or models, according to OMG[18] or Booch *et al.* (1999), *"every interface should represent a seam in the system, separating specification from implementation"*. Let us call it *OMG-interface* to distinguish it from its typical meaning in CAx-context.

A system that possessing interfaces is sometimes called an *open system*. Theoretically it is also possible to have the opposite type of system – a *closed system*, but since such systems cannot communicate with the outside world, they are not of interest for our study (except if we are inside such a system), and are not discussed further.

The systems can be classified according to different criteria. An example classification is given in Figure 2.39.

2.4.2.3.2.5 Systems Engineering

Systems engineering (also known as *systems design engineering*) is a relatively new (originating around the time of World War II) branch of the science with focus on the definition, realization and characterization of complex, but at the same time successful systems. Some of the well-known subfields of systems engineering are safety engineering, reliability engineering, interface design, cognitive systems engineering, communication protocols, security engineering.

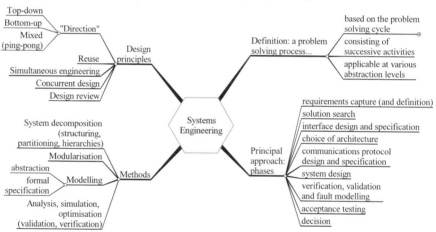

Figure 2.38. Inherencies of the systems engineering

2.4.2.3.2.6 Software Model vs. Computer Model

The terms software models and computer models are often used. Before we make use of them, we shall clarify what is similar and what is different between them.

A *software model* is an implementation of an information model (*cf.* Figure 2.5). It is a kind of representation (or mapping) of the information model by means of data structures and algorithms. Typically some formal languages are used to code the algorithms and the respective data structures into *programs*.

---

[18] Object Management Group, Inc. *Cf.* http://www.omg.org/

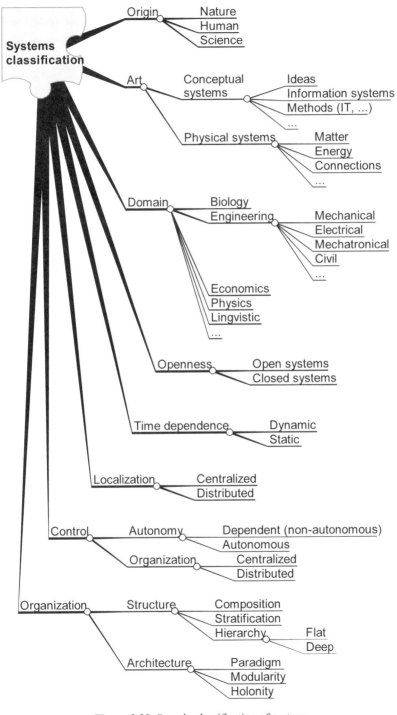

Figure 2.39. Sample classification of systems

Therefore, any software model can be viewed as a set of programs and data. Software models can be kept or transported on a medium, but the medium alone is not enough to allow the model to "come to life". This can be achieved only when the model is loaded in some hardware, and the control is transferred to it. Thus, we can say that the model "lives" in its host – typically, a computer.

Without software no computer can directly be used to represent anything. When we say that something is "modelled with computer" we mean: "modelled by means of software running on a computer". Thus, a *computer model* is nothing else than a software model that is loaded into a computer and activated there. In other words, both terms can be viewed as synonyms referring to a system consisting of hardware, programs and data. The main difference is that each of the terms stresses a different aspect of this kind of modelling, *e.g.* the use of software or the use of hardware, with the respective specificities.

2.4.2.3.2.7 Computer Aided Systems

A Computer Aided system or in short *CA-system* (also spelled without dash) is a complex software system, dedicated to solving tasks in a specific subject area. The subject area is typically a phase of the product lifecycle (design, planning, manufacturing, marketing, *etc.*) or an activity existing in many phases (*e.g.*, quality control, product-data management) and its name is usually reflected in both the long and the short forms (*e.g.*, Computer Aided Manufacturing system or CAM system). The alternative short form *CAx-system* is often used as a collective name for all possible short forms, where "x" is a placeholder, matching the name of any phase or activity. The "computer aided" is not really an obligatory part of the name, since it is implied for numerous activities. Thus, nobody would speak about word-processing without some kind of computer, but the respective "computer aided" system – the word processor or word-processing system – meets the definition and should be considered as belonging to the group of CAx-systems, too.

When several CAx-systems are used to automate related activities and are developed from the same producer they are often referred to as *software packages*, *software packets* or *suites*.

2.4.2.3.2.7.1 CAx-model

The usage of CAx-systems has become so common during recent decades that many people tend to forget: most CAx-systems create as one of their outputs a model (CAD-model, DMU-model, FEM-model, *etc.*). This model is often either the most important or the only result produced.

Some of these CAx-models are product models, some of them are object models (*i.e.*, something that is not going to be produced, but is used as part of other models) and some are models of processes. Therefore, the term CAx-model is used in the text as a generic term, referring to models of any of the types mentioned.

2.4.2.3.2.7.2 CAx-system Centred Approach

For each product several different product models, related to different phases or aspects of its lifecycle, are created and used. Typically, the model related to a given phase or aspect is prepared by a dedicated (CAx) system and can be modified and further developed only by *identical* (*i.e.* having the same type/dedication *and* from the same producer) or *compatible* (*i.e.* capable to read the models in their initial format) system. Even more important is that the models, created from a CAx-system, can be used mainly *within* the system-creator, sometimes – *within* another (foreign) system and almost never – *alone*. In other words, any model "lives" only within a given CAx-system, and it in turn "lives" only within the respective hardware. For these reasons I shall use the name *system centred approach* (in short, *SCA*). The system where a specific model lives is called its *host* or *host system*.

In contrast to the object-oriented approach, where everything is centred around the concept of an *object*, and where the objects' *methods* (algorithmic description of operations on objects) are typically defined only on the objects of the same class (*i.e.* type), CAx-systems operate on *diverse types of objects* but can perform only a given *group* (or *class*) *of operations*[19]. Thus, a general term for referring to all types of CAx-systems together could be "operation-oriented system". Since in the computerized support of production we more often speak about classes of operations having to be performed than about classes of objects having to be processed, we have one more reason to say that conventional computer aided production is *system centred*.

2.4.2.3.2.8   Data Integration
The process of data integration includes collecting all data "pieces", putting them in the same (or in compatible) format and possibly performing other actions to ensure that they are usable together.

2.4.2.3.2.8.1   Data Transfer, Data Exchange
The term data transfer/exchange refers to all actions that have to be performed in order to make a model created or existing on a given platform, work on another platform. These actions include the physical transfer of the model and possibly conversions (translation) on different levels. One could speak of unidirectional and bi-directional exchange, as well as exchange among multiple platforms. Typically, data exchange uses a file as entity, is unidirectional (*i.e.* it is transfer and not real exchange!), occurs offline and not so often as the data processing itself.

2.4.2.3.2.8.1.1   Models of the Data Exchange and their Qualities
As any other process, the process of data transfer or exchange can also be modelled. The resulting models can be divided into two main groups, aiming either at process development and realization or at its analysis and possibly – requests for improvement. A short classification of data exchange is given in Figure 2.40; more different models and a discussion of their qualities can be found in Avgoustinov (1997).

2.4.2.3.2.8.1.2   "Interface Pressure"
Suppose that a "force" called *demand for data exchange* exists, and that the *descriptive potential*, defined in Avgoustinov (1997) as the cardinality of the set of elements of the source language, symbolizes the "area" on which the force is applied. Then we could use the metaphor "interface pressure", which (exactly as normal pressure) is proportional to the force and inversely proportional to the area.

---

[19] This does not mean that the object-oriented approach is not used in CAx-systems; it is simply applied on a different (lower) level.

Similarly, it is possible to say that the target system has "*interface resistance*" that is also proportional to the demand for exchange, and inversely proportional to that part of the descriptive potential which is utilized in the models to be transferred.

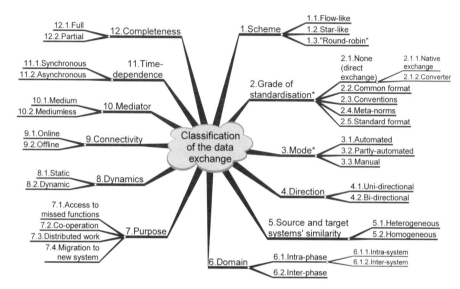

Figure 2.40. An example classification of the data exchange. Reworked and extended, after Avgoustinov (1997)[20]

2.4.2.3.2.8.2 Data Sharing

Data sharing is a general term used to denote the process of making data available to more than one user or system. In contrast to data exchange there is no typical entity. Only needed data is accessed and the sharing occurs on demand, online, multidirectional and even multiple times per processing. Since typically only very small parts of the source model are needed, accessing only them and only on demand makes the process much more efficient than data exchange.

2.4.2.3.2.9 Component

We shall refer to any at least logically separable part of a model, product, system or any other compound object as a *component*. When a component is not compound, we can call it also an *element*. In the domain of software programs or models, according to OMG as in Booch *et al.* (1999), a component is "*A physical and replaceable part of a system that conforms to and provides the realization of a set of interfaces.*"

---

[20] Slightly different notation is used in the cited publication for the two terms, denoted "*" in Figure 2.40, but the actualized notation (as given in the figure) seem more adequate to me.

2.4.2.3.2.9.1 Integration of Components

On the one hand, components (both software and physical) are designed to be integrated into systems and also to be interchangeable. On the other hand, they cannot be combined arbitrarily. The process of achieving effective interaction (including communication) among components, as well as interaction of the user with these components independently of their location, host, state, *etc.*, is called in this study *component integration*.

2.4.2.3.2.9.2 Component-based CAx-system

Any CAx-system that is built up from software components is called *component-based CAx-system*. According to Kilb and Arnold (1998) *"... using a system based on the CAx object bus, there is no longer a need for file based data transfer between different integrated systems. Only the necessary representation information of the data models has to be transmitted as CAx objects through the CAx object bus."* The problem here is that the "necessary representation information" has not only to be transmitted but – depending on the case – also converted to the respective format!

2.4.2.3.3 Relation to Suitability, Relevance, Adequacy, Reusability

The components and their organization and granularity have a strong influence on many traits of the system that contains them. Below are enumerated some relations of the components (or componentization) to some of these traits or to other notions of interest; in all formulas $f(...)$ symbolizes a function of something, whereas F denotes some kind of functionality (*e.g.*, a set of functions), explained by an appropriate subscript; after some formulas are given their ranges.

$$Suitability = f(form,\ granularity) \tag{2.17}$$

If all functions composing the functionality are equally important and their implementation is either full or null, suitability can be expressed as the ratio between the cardinalities of the set of (fully) covered functions and the set of needed functions.

$$Suitability = |F_{covered}| / |F_{needed}| \qquad [0,\ 1] \tag{2.18}$$

Since usually some functions are more important than others, the above formula should be extended with weight factors for each functions. To keep results within a normalized range (*i.e.*, within [0,1]) the sum of all weight factors should be equal to 1 (or 100%).

$$Suitability = f((coverage - needs)/needs) \qquad [0,\ 1] \tag{2.19}$$

The gaps between the granules (*i.e.*, the uncovered areas) decrease the suitability and together with the excess functionality, form the inefficiency of the system.

$$Adequacy = f(overlaps,\ shortage,\ excess) \qquad [0,\ 1] \tag{2.20}$$

It can be reasonable to use the inadequacy instead:

$$Inadequacy = f((overlaps + shortage + excess)\ /\ needs) \tag{2.21}$$

The reserve (or spare) functionality can be expressed as follows:

$$F_{Reserve} = F_{total} - F_{needed} \hspace{3cm} (2.22)$$

Analogously, the redundant (or excess) functionality can be expressed as a function of the doubled (or overlaped) in different components functionality:

$$F_{redundant} = f(F_{overlaped}, F_{needed}) \hspace{2cm} (2.23)$$

2.5   Model Representation

After its elaboration, each idea has to be represented somehow. The representation allows us to communicate the model to others, to save it for later use and even to discover things or relations among them that were invisible or not obvious before. When the modellee is not an idea, but something really existing, the situation is not very different: the main difference is that the modeller first "gets the idea" and then prepares a representation, reflecting the most important and relevant-for-the-purpose properties of the modellee.

2.5.1   Reasons for Discussing the Representation

Apart from the fact that for software models the representation has great influence on the efficiency, compactness and other characteristics of the model, there are other reasons for discussing the models' representation:

- very often models are mixed up with one of their representations;
- improper representation could lead to confusion or loss of information;
- some models represent not a single entity/modellee, but a whole class of similar entities (modellees); we shall call such models *parameterized*;
- the model's representation is non-abstract (real) in contrast to any software model itself.

Very often models have more than one representation – *e.g.*, if the same model were drawn twice, but in two different colours, we would get two representations of the same model. But we should not confuse different models of the same modellee with different representations of the same model: if we represent some modellee once as a text and once as a drawing, these would be two different models of the same modellee.

On the other hand, some models can have different views or aspects – for instance, a three-dimensional (3D) model can be viewed from different viewpoints; although each viewpoint can be represented on paper as a 2D drawing or snapshot, they remain different 2D representations of the same 3D model.

2.5.2   Classification of the Representation Types

The representation of a model can depend on many different things – medium, method, changeability, dynamics and others. The representation can even change during the model's use. Since each representation type has its advantages and disadvantages, the modeller can choose the type most appropriate for the purpose. Therefore, we say that each representation is purpose-dependent. Sometimes –

especially with respect to multi-purpose models – the modeller can choose to provide several representations, so that the user of the model is able to choose the most appropriate for the moment or for the respective task representation.

We shall distinguish between two different types of model representation: internal and external.

The *internal representation* concerns how a given model is represented within a software system or within a computer, and is implementation dependent. For instance, the internal representation of the simplified model of a circle from Figure 2.17 can be as short as three numbers, connected with the knowledge that they represent the radius and the coordinates of the centre point, respectively.

The *external* (or *observable*) representation of a software model is usually a dynamic representation, depending on the values of the model data at the moment. Figure 2.17 itself is an external representation of the (parameterized) circle. It should be noted that the external representation is usually based on the internal one.

Within the internal representation we distinguish between *model data* (or parameters) and *model invariance* or *model knowledge*.

The model data is different among the instances of the modelled class of entities and is used to create distinguishable representatives of the class. It is almost always included in the model saved on a medium to guarantee its persistence.

Model invariance can be of two subtypes: *programs* and *metadata*. A program can be viewed as data, describing one or more algorithms. It can describe operations on real data or on placeholders. At runtime the placeholders are replaced with the actual values of the parameters.

The metadata describes the relation among the data elements at the lowest level (the parameters) and is usually implemented by means of data structures.

On the next level the relations among the metadata can be described by means of meta-metadata – *cf.* Figure 2.17 once more. Since we can always describe the relations among elements of one level by means of *$meta^x$-data* on the next higher level, we can speak about *metadata of different degree* (*cf.* Section 2.4.2.1.2.3.1 Levels of Hierarchical Structure above).

2.6   Integration of Models

The majority of devices, machines and other products are actually complex systems built up from separately produced components. These components can be of pure mechanical, electrical, or electronic nature, or they can also be intermixed. When users observe and use them as a whole – *i.e.* the product – there is no need to speak about integration from the user's point of view. From the manufacturer's point of view, however, all components have to be assembled or built together; this process can be viewed as integration. Therefore, when a compound product is modelled, depending on the purpose of the model, it could be natural and useful first to model each component alone and after that to integrate all these models in a compound model of the product. But what is integration? According to Lutters (2001), there are countless definitions of integration. One more reason to define again what should be understood under integration in the present work is that none of the known definitions is perfect, including this in Lutters (2001): "*the facilitation of mutual cooperation and interaction between distinct functions in the manufacturing environment*". Weak points in the last definition are the lack of an aim in the definition and the word "facilitation"; a better attempt would be to use "accomplishing" instead. Our approximation for a more general definition could be the following:

Definition 2.12: The integration of two or more (manufacturing) components is the process of making them work on one and the same task or contribute to achieving one and the same outcome.

How exactly integration will be achieved – whether the components will be "physically joined" or "obey the same control", or just the results of their work will be joined – is a question of secondary importance. In other words, the integration is not an end in itself: either the result, achieved after the integration of two models, is better than the sum of the results of the two non-integrated models, or such a result cannot be achieved at all without integration.

In the simplest case integration of two models would be to achieve their simultaneous use within (or from within) the same environment (hardware, operating system, application software, *etc.*). For more sophisticated dynamic systems, though, making the communication between the models involved and the interaction (of the user) with them effective, independently of model location, host, state, *etc.*, is also indispensable.

2.6.1 Integration Classification

The integration can be classified according to different criteria. Perhaps the most important criterion is the type of components that have to be integrated – real or virtual, material or abstract, *etc.* – since this can affect most of the other criteria. It is apparent that the method for integration of the components of a real car will be different from the methods for integration of the models of the same car components.

The technique used for achieving the integration can serve also as a characteristic of classification of integration techniques. An example is given in Figure 2.41.

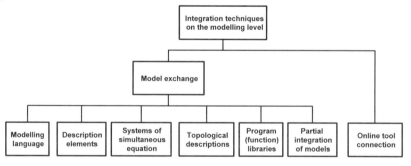

Figure 2.41. Some integration techniques, after Gausemeier and Lückel (2000)

A sample classification according to several additional criteria (inherent integration traits) is given in Figure 2.42.

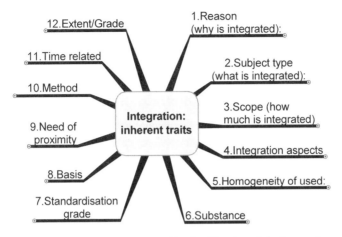

Figure 2.42. Example for a possible classification of the integration

2.6.2   Models, (Software) Applications and their Integration

Computer programs (a.k.a. software programs) that aim to solve user-specific (and not system-specific!) problems are often called *software applications*. They reside on top of the operating system and are very often used to:

14. control some local-computer-related process like printing, visualization, communication, *etc.*;

15. control some process that is not related to the host computer, but to business or some other part of real life – production, transportation, commerce, *etc.*;

16. model and simulate different real processes;

17. support various processes otherwise.

Exactly like many other things that have structure or are part of a structure, applications may need to be integrated. Even if we consider only case 16 above, models of sub-processes have to be integrated to achieve a simulation of the full process that is modelled. In the case of modelling, simulation or control of complex processes it can be necessary to integrate many applications of different type, different origin, and different sites within a given enterprise and even located in different enterprises. In similar cases an often used term is *enterprise application integration*. One of the popular definitions is given in Wikipedia Wiki (2006):

> *Enterprise application integration (EAI) is the use of software and architectural principles to bring together (integrate) a set of enterprise computer applications. It is an area of computer systems architecture that gained wide recognition from about 2004 onwards. EAI is related to middleware technologies such as message-oriented middleware MOM, and data representation technologies such as XML. Newer EAI technologies involve using web services as part of service-oriented architecture as a means of integration.*

Achieving integration – or even better, integrability – is often more important than achieving good coverage or good functionality. Again in Wikipedia Wiki (2006), the following is said about the role of the integration:

> *Without integration, enterprise computing often takes the form of islands of automation, where the value of individual systems is not maximised because they are working in partial or full isolation. However, if integration is carried out without following a structured EAI approach, many point-to-point connections grow up across an organization. Dependencies are added on an ad hoc basis, resulting in a tangled non-maintainable mess, commonly referred to as spaghetti.*
>
> *...*
>
> *EAI is not just about sharing data between applications. EAI focuses on sharing both business data and business process.*

Many different integration approaches have been developed and tested, achieving great or small success, but none of them has been generally accepted. One of the objectives of this book is to clarify the role of the modelling approach in achieving satisfactory model integration and therewith also better integration of the respective modellees. More details are discussed in the following chapters.

3

Conventional Product and Process Modelling

In all science, error precedes the truth, and it is better it should go first than last.

Hugh Walpole

Although many of the findings and considerations in this chapter could be generally valid for any kind of computer-based modelling, they are based mainly on observations of and experience with computer-based modelling in the field of *mechanical engineering* and *mechatronics* (MEM-modelling).

3.1 Problems of Contemporary Modelling

Understanding the problem is more important than finding a solution, since the exact representation of the problem automatically leads to the proper solution.

Albert Einstein[21]

As with any other area of science and research, contemporary modelling has its problems and weak points. This section presents an analysis and discussion of some important problems in contemporary modelling, in an attempt to extend or improve their understanding, as well as to propose some improvements. It starts with some more general observations, thereafter some of the more specific issues are discussed as well.

---

[21] *"Das Problem kennen ist wichtiger, als die Lösung zu finden, denn die genaue Darstellung des Problems führt automatisch zur richtigen Lösung."* (attributed to Albert Einstein by Adrian Krahn in "Vom Prozessmonitoring zum Prozessmanagement").

3.1.1   General Observations

> *I keep six honest serving-men*
> *(They taught me all I knew);*
> *Their names are What and Why and When*
> *And How and Where and Who.*

Rudyard Kipling
The Elephant's Child, 1902

Let us start with enumeration of some important facts and observations, evaluating their influence on the modelling domain, and then attempt to find their causes and consider how their negative influence could be avoided or relieved. Many of the phenomena that cluster together are so interdependent that it is not trivial to choose the (right) sequence for reviewing the elements of such a grouping. For this reason, as the discussion progresses and new information is presented, some of the points will be addressed again and again.

3.1.1.1   The Most Popular Modelling Approach
During the last twenty years, despite numerous alternatives, the use of software models on a computer has gradually become the most popular modelling approach. Why is this approach gaining more and more success and acceptance, despite the fact that it is the youngest and that at the beginning of the computer era both computers and software were much more expensive than nowadays? Indeed, this phenomenon is observable in almost all branches of science, industry and social life. However, the explanation should be sought in the character of modelling itself, rather than in the specificity of any particular area.

An estimation of how suitable the model medium (or model nature, *cf.* Figure 2.26) tends to be for fulfilling general modelling purposes is given in Table 3.1, where the following assumptions are made:

- The nature of the model medium can vary for the same modellee.
- All estimations consider theoretical possibilities rather than implementation-dependent values.

The averaged results presented in the table are estimated on the basis of considering a number of different modellees from different domains.

As we can see from this table, the use of computer models always leads to the best fulfilment of common purposes of the modelling. But is the price acceptable and do we always get only advantages, compared to other natures of model medium? Some other (mainly economic) considerations are given in Table 3.2.

In Table 3.1, Table 3.2 and Table 3.3 the given values are relative to other estimations, where the following notation is used:

- the maximum of the respective attribute for all media is denoted *max*;
- the minimum is denoted *min*;
- absence of an attribute is denoted "-";
- more than one alternatives are separated with "/";
- a partial presence of an attribute is denoted "~";
- more "+" symbols mean a stronger relation or better representation of the respective attribute.

Table 3.1. Fulfilling the general modelling purposes depending on the model medium

Purpose \ Nature/medium	Durable material	Workable material	Imagination	Mathematical model	Software or computer model
Supporting and improving the understanding of the matter	++	++	+	++	+++
Supporting communication	+	+	-	++	+++
Providing a common basis for discussions and information exchange	yes	yes	no	yes	yes
Allowing comparison of different solutions	+	+	~	+++	+++
Allowing analysis and prediction of characteristics	+	+	-/+	+++	+++
Allowing analysis and prediction of behaviour	+	+	-/+	+++	+++

Table 3.2. Estimation of modelling properties depending on the model medium (a low value is more advantageous!)

Modelling qualities \ Nature/medium	Durable material	Workable material	Imagination	Mathematical model	Software or computer
Modelling price			min		max
Time needed for modelling			min		
Model price			min		
Price of model exploitation			min		
Price of duplication	max		n.a.	*	min
Time needed for duplication	max	**	n.a.	*	min
Price for model transportation	max	**	n.a.	*	min
Price of the host of the model	n.a.	n.a.	n.a.	n.a.	max

And finally, we shall consider not only the qualities of the modelling, but also the qualities of the results, *i.e.* of the resulting models. In Table 3.3, only qualities that are modellee-independent and applicable to any nature of model medium are listed.

Table 3.3. Estimation of different model properties depending on the model medium

Model qualities \ Nature/medium	Durable material	Workable material	Imagination	Mathematical model	Software or computer
Reusability	+	++	-	+++	max
Autonomy	n.a.	n.a.	n.a.	n.a.	max
Changeability	+	++	max	+++	+++
Complexity			?	+++	max
Dynamics	min		max		
Durability	max		min	max*	
Extendibility	min	++	max	+++	+++
Flexibility	min		max		
Homogeneity			max		
Intelligence	n.a.	n.a.	n.a.	n.a.	max
Independence	max	max			
Openness and modifiability	min	+	max	+++	+++
Reliability					
Robustness					
Scalability					yes
Universality	min	+	max	++	+++
Updateability	min	+	max	++	+++
Remote usability	n.a.	n.a.	n.a.	n.a.	yes

3.1.1.2 Dependency on the Achievements of the Information Technology

Computer modelling has its disadvantages, too. Most of them stem from the fact that any computer model is a heterogeneous system of hardware, programs (software) and data. Thus, computer modelling depends on the achievements of at least three different branches of science and technology: hardware engineering, software engineering and computer science (a.k.a. informatics).

Brooks Jr. (1987) classified the difficulties of software technology in two categories: "...*essence, the difficulties inherent in the nature of software, and accidents, those difficulties that today attend its production but are not inherent.*". Also there has been mentioned: "*as we look to the horizon of a decade hence, we see no silver bullet.*" Now, almost two decades later, we can say that this forecast has become true: although many novel methods, tools and improvements were introduced in software engineering, none of them was a panacea (a universal and perfect solution) to the existing problems and difficulties. In particular, the accidental problems of software modelling are tightly bound nowadays to the three branches mentioned above – software engineering, hardware engineering and informatics.

3.1.1.3 System Centricity of the Conventional (Approach to) Modelling

At the beginning of the 21st century, the modelling of products and processes is still performed mostly by means of computer-aided systems, primarily for historical reasons. The trend began with the spread of formal programming languages (which corresponds approximately to the second stage in Figure 3.1) in the late 1950s and especially with the idea of writing each new routine in a way allowing for its reuse (*cf.* the third stage in Figure 3.1). Such parameterized routines were collected into libraries and each newly written program was able to simply call them and rely on their functionality. A higher number of routines in the library used to mean higher functionality, therefore there was a naive striving to make the libraries larger and to write reusable programs. With the growth of routine diversity in the libraries, one began to lose track of what was available, so a new trend emerged for reorganization of the routines and grouping them in purpose-oriented libraries. Then a new type of programs was due, that would allow use of as many of the library routines as possible with as little effort as possible. These new program types were initially called applications (or application software). The latter were also known under different domain- and purpose-dependent names (*e.g.*, word processor, spread-sheet, plotting program, *etc.*), but were actually the first computer aided software systems, known under the short name *CAx-systems*.

With time, the use of CAx-systems as modelling tools became the dominant approach to development of complex models – at least in the field of mechanical and electronic engineering. We shall call this approach the *system centred approach* (**SCA**), or *conventional approach* to modelling. For about a decade or two (say, between 1975 and 1995) the CAx-systems seemed irreplaceable and untouchable aid in the engineering processes, and especially for engineering modelling. However, the continuously increasing number of types of CAx-systems (continuing even nowadays) started to produce software systems deviating from the "typical CAx-system"– *i.e.* distributed systems, intelligent systems, *etc.* A common characteristic of all CAx-systems, though, remains the permanently increasing number of suppliers that are involved in their production. This situation is relatively easy to explain: permanently increasing requirements lead to higher required functionality, which, in turn, means involvement of more and more domains. Consequently, teams developing a given CAx-system involve more and more experts, which – due to specialization – (have to) come from more and more different enterprises.

It is highly probable that in the future the focus will be shifted from CAx-systems, which are actually only tools helping us to build models, towards the models/results themselves, which are what we need. Thus, it is not impossible that the time of The System comprised of many autonomous intelligent models (*cf.* the last stage in Figure 3.1) will become reality. Until then, the SCA remains a dominating approach, as it has many undisputable advantages. But since it has some peculiarities, too, at least the most important of them will be discussed in dedicated sections below.

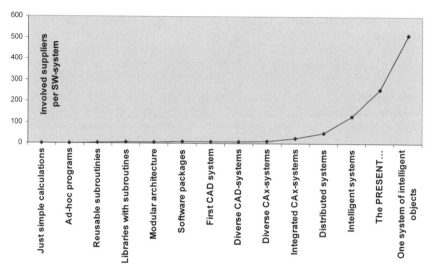

Figure 3.1. Some more important achievements in the CAE-history, after Avgoustinov and Bley (2006)

3.1.1.4 Communication

Recent globalization has led to increased competition in the market. In order to remain competitive, manufacturers can try to increase the quality, to decrease the manufacturing costs, to reduce the time-to-market, or to play with all of these factors. Specialization is a way to achieve better quality and possibly lower price, but it leads to inability to accomplish all tasks, so it should be combined with cooperation for influencing both price and quality. Cooperation, in turn, could lead to additional cost reduction (*e.g.* achieved by outsourcing), but leads to urgent need for global data access and communication in the form of **data exchange** and/or **data sharing** among the involved software systems. According to Sachers (2001), the former is characterized by a file-based, asynchronous data exchange between two partners, while the latter represents "*a synchronous data access using a network. Data will be not replicated. There can be links between data objects across the network. For the user it should be transparent where the data are physically stored.*" It is commonly agreed that both types of communication – *i.e.* data exchange and data sharing – are necessary, but depending on the application one of them could be more adequate. Since each cooperation partner is typically in a different location, uses a different platform and has different specialization and tasks, the resulting cooperation environment is heterogeneous and multicultural. It is hardly surprising, therefore that despite numerous standards for information exchange, communication among partners and integration of the respective (model) pieces often encounter problems.

3.1.1.5 Security

Communication provides means for cooperation during the definition, design, planning, production, use and support of different products in/from different places. Yet, communication via public lines (and especially via Internet) raises many security issues. On the one hand each partner should be secured against intrusion and data theft, which is achieved through password protection and firewalls. On the other hand, the communication line has to be protected against tapping from third parties, which is accomplished by data encoding. For some manufacturers all this is insufficient, thus, the European Network eXchange (ENX) was founded in 1997 – a network with no direct link to public networks (to ensure security), devoted to cooperation among participants. With or without an ENX-like network, there is still a problem here, which comes from "inside". In order to cooperate with each other the participants have to communicate, so they have some access to their partners' net or computer. But to allow every partner to cooperate without having to expose sensitive know-how to other collaborators, a mechanism for fine-grained access permissions is needed that is not always available. Ideally, access permissions can be defined differently for each level of any model and for each component within a level and can be based on account or user group, access time, user location, *etc.* From another point of view, many authors (or their enterprises) could be interested in access protection that allows the partners to use the respective object (model, product, *etc.*) but at the same time protects the author's know-how from theft or undesired revealing.

3.1.1.6 Integration-related Issues

The integration of both separately developed products and their (separately developed) models has remained for decades one of the burning issues of contemporary engineering. It has affected not only the sub-domains of mechatronics, but all branches of science, industry and public life. Despite immense investments in standardization (*cf.* Section 3.1.1.12) aiming at improvement of the integrability, the results are moderate. Due to the importance of integration related issues, they are discussed in a dedicated section (Section 3.1.3) below.

3.1.1.7 Enterprise-related Observations

3.1.1.7.1 Specificities of Small and Medium-sized Enterprises (SMEs)

According to Günterberg and Kayser (2004), 99.7% of all enterprises subject to VAT in Germany in year 2003 were SMEs. They employed 70.2% of all employees in private businesses and realized 41.2% of all turnovers subject to VAT. In 2003 the Commission of the European Union European Union (2003, ANNEX, Article 2) changed the definition for SMEs, with validity from the beginning of 2005, as follows:

> *The category of micro, small and medium-sized enterprises (SMEs) is made up of enterprises which employ fewer than 250 persons and which have an annual turnover not exceeding EUR 50 million, and/or an annual balance sheet total not exceeding EUR 43 million.*

According to another evaluation of the statistical data about Germany in 2003, obeying the new definition of the EU, 99.6% of all enterprises were SMEs, they employed 55.5% of all employees and realized 40.9% of the turnover.

Since the distribution of SMEs in other countries of the European Union (and, in all probability, in most other industrial countries) is not very different, they deserve special attention in the analysis.

The SMEs have some specificities. They usually possess knowledge in a relatively small area, but this knowledge is very detailed. They have restricted resources and irregular customers as well as often changing tasks. Consequently, they cannot accomplish every task alone and seek cooperation with other enterprises. The chances to find appropriate partner(s) in the neighbourhood are relatively low. In addition, a single partner is rarely sufficient. So, in most cases a SME has to cooperate with numerous and geographically dispersed partners.

The mentioned specificities leave us with the impression that even minimal improvements of the following aspects would have an immense impact on the industry as a whole and are welcome:

- Tools for optimal representation and exploitation of the available knowledge and expertise.
- Possibility to use cheap, reusable components.
- Maximal flexibility and reusability of the employed concepts, tools, components and other resources.
- Tools supporting cooperation, in particular, tools supporting data exchange and/or integration with foreign models.
- Reliable communications and net-based solutions.

All these considerations are summarized in Table 3.4. In addition, it should be noted that since the end of the 20th century there has been an obvious trend to substitute expensive products or services with services on demand over the Internet under the pay-by-use scheme.

Table 3.4. SME-related observations and their consequences

Observations	Consequences/requirements
Have detailed knowledge in a relatively small area	Tools for optimal representation and exploitation of this knowledge;
Cannot accomplish everything alone	Cooperation $\Rightarrow$ data exchange or integration with foreign models;
Have restricted resources	Cheap, reusable components; Leasing or pay-by-use instead of purchase; Outsourcing;
Often changing customers and tasks	Maximal flexibility and reusability;
Have numerous and geographically dispersed partners	Reliable communications and net-based solutions.

3.1.1.7.2 Enterprise Uniqueness

There are at least two cases of enterprise-specific modelling, namely, modelling of products manufactured in the respective enterprise and modelling of the relevant processes. The former case is meaningful mainly if the modelled product is unique and produced nowhere else. But even then its importance is much lower than that of the latter case. The reason is that processes are modelled in order to be optimized, and they either include products or influence them. Furthermore, to model an enterprise-specific process often means to model the enterprise itself. Since two enterprises are never the same, their (digital) models will always be different. This means that models of enterprises have almost no chance to be reusable, which makes the modelling rather expensive. On the other hand, it is much more often the case that certain parts (components) of enterprises happen to be the same or at least similar. So, the components of an enterprise model can be developed as parameterized and reusable components. Nevertheless, current modelling approaches do not offer a satisfactory solution for the creation of components of process models which could be reused without exposing enterprise-specific know-how or secrets.

3.1.1.7.3 Influence of the Globalization
Last but not least, let us recall that the globalization poses specific requirements and raises important issues. Among them are the severe competition in price, quality and delivery speed; the cooperation among partners in different countries and time zones; the need to integrate parts or components created by a multi-cultural society into functional and reliable products, as well as the ability to organize the production, distribution and maintenance of products in huge number of variants and for a large number of highly diverse conditions of use (climate, standards, erudition of the average user, *etc.*).

3.1.1.8 Lifetime
The lifetime of a model is one of its important traits. It seems to be branch-specific and product-specific and there is no common opinion on its minimal length. Yet, typical lifetime is between 3 and 5 years for hardware, about 8–10 years for system software, 15–20 years for applications and about 30 years for data/information. In particular, "*25 years as recommended for the automotive industry*" is mentioned in Kilb and Arnold (1998), and the desired minimal lifetime in the aerospace industry is longer – a frequently mentioned number is fifty years, and there is a trend for requiring even longer lifetime for data and its derivatives (information, knowledge, *etc.*).

We face an obvious paradox here. On the one hand, a long model/data life is pursued. On the other hand, the models are dependent on the CAx-system (application). Apparently, the simpler the format of the data, the more rarely a need to change it arises and, respectively, the longer the life of the data represented in that format. A decisive factor for the lifespan of applications could be their popularity, and for data – the popularity of the format.

For example, Burkett (1998) aims at "*data usage architecture that starts with the assumption of data access and exchange over the Internet using existing Internet protocols and languages. This provides a general and ubiquitous platform that will be more stable and have a longer lifespan than point-specific solutions.*" But the data exchange forms a separate group of issues. Yet more interesting and important is to consider the lifetime of processes and their models.

Some generalized process types (*e.g.*, forging) are as old as the manufacturing itself and are nevertheless (with respective modifications and improvements) still in use. It would be advantageous if models of any process can live at least as long as the process itself. The more general a given process is, the more general (and simple) its model will be. So, the lifetime of a general process and its model is expected to be respectively longer than that of more specific processes and their models.

3.1.1.9 Changeability and its Derivatives

The desire for increased lifetime of most products as well as for increased lifetime of (the respective) production facilities poses new requirements on the manufacturing process, on its planning, organization and control. One of these requirements is the ability to adapt easily and quickly to new requirements.

According to Wiendahl (2002), *"The changeability of enterprises develops more and more to the key success factor, in particular because of new conditions in the surrounding field of production."*. In particular, Wiendahl distinguishes five levels of changeability, viewing them in two orthogonal dimensions – the product level and the production level (*cf.* Figure 3.2). On the lowest level, he introduces the *change-over ability*, which is relevant to the operation and to a workplace and is defined in Wiendahl and Heger (2003) and Wiendahl and Heger (2004) as *"the technical ability of a single machine or workstation to perform particular operations ... with minimal effort and delay"*.

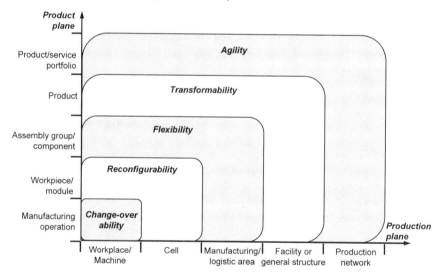

Figure 3.2. Changeability and its derivatives, after Wiendahl (2002)

Again, at the second level we have the *reconfigurability*, which corresponds to a (product-) part and to a manufacturing cell and describes the ability to switch with minimal effort and delay to production with other characteristics.

On the third level, the ability to make changes in an assembly group and, respectively, in a manufacturing area is referred to as *flexibility*. It is viewed as the easiness of switching the production to other – although similar – product (family).

The ability to make changes in a product and, respectively, in a facility on the fourth level is referred to as *transformability* and indicates the ability of a whole facility (*e.g.*, factory) to switch to the manufacturing of another product family.

Finally, on the fifth level, he uses the notion of *agility* to describe the strategic ability of an enterprise to open up new market(s) and undertake every necessary measure for adapting its product/service portfolio and its production network to it.

All these types or derivatives of changeability hold to some extent for models and modelling, too. With software models, though, it makes sense to introduce one additional (lowest) level – the level of variables or, respectively, modelled attributes. I shall call the possibility to make changes on this lowest level simply *changeability*. The relations among the levels of modelling and the derivatives of changeability according to the idea of Wiendahl (*cf.* Wiendahl (2002, p. 126) are represented in Figure 3.3.

Note that some elements of the model level appear on the boundary between two changeability levels, which means that depending on the case the one or the other changeability types could apply. This should not be viewed as a contradiction or exception, but – instead – as a confirmation that the software is, in general, more changeable/flexible than anything else.

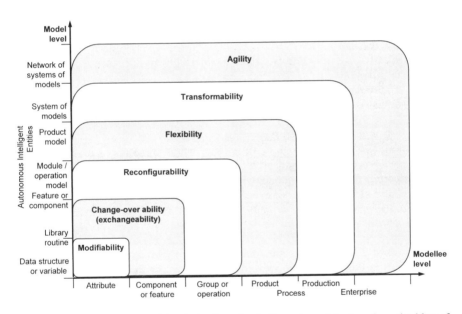

Figure 3.3. Changeability and its derivatives for (software) models, based on the idea of Wiendahl (2002)

Another fact to bare in mind is that in the area of software engineering and software modelling the term flexibility seems to be used as a generic replacement of all changeability derivatives. One of the reasons is that the term changeability is neutral – *i.e.*, it can be used in both positive or negative sense, since – depending on the case – changeability can be desired or not desired[22]. Flexibility, on the other hand, is typically used to describe a desired trait and is usually related to

---

[22] For instance, the changing (or changeability) of a read-only variable/model/document, *etc.*, is usually not desired.

adaptability (*cf.* its definition in Section 2.4 above in the previous chapter). Another probable reason is that the term is somewhat related to the term software – via the word(part) soft – and probably sounds more familiar.

Although I adhered to the original terms in Figure 3.3 (except for the lowest level, which is newly introduced), in the case of software models they are not always appropriate. The computer scientists use predominantly the term *flexibility* on different levels, although there seems to exist no common view on the changeability in respect of structure and hierarchy of sophisticated models/systems. This is reflected in the definition of flexibility from IEEE (1991): *"the ease with which a system or component can be modified for use in applications or environments other than those for which it was specifically designed"*. Therefore, instead of (re)defining or explaining all these terms for software, I would like just to note that the last three of them – flexibility, transformability and agility – actually describe *the ease with which a modellee on a given level or its corresponding model can be adapted to new needs, requirements or purposes.* Since this resembles the definition of IEEE fairly close, I prefer to refer to the whole group of terms as *levels of flexibility* or *derivatives of flexibility*.

Please note that due to the attempt to keep general validity on many places in both Figure 3.2 and Figure 3.3, it is possible to use several alternative terms, but due to lack of space only one or two of them are given. For instance, the place of the *network of systems of models* could be taken from *distributes system of models*, instead of *product model* on the same level could stay *model of a system* (of objects) and so on.

3.1.1.10 Mechatronics-specific Requirements/Issues

Mechatronics is a typical example of an area where three major branches merge into one: mechanical engineering, electronics and software engineering. Consequently, the product models have minimum three layers – mechanical, electrical and software – but not every CAx-system offers support for all of them. Although the integration of components from these three layers is indispensable, it is often problematic due to irresolvable differences (format, representation, conventions, *etc.*) between the layers of the model or due to incompatibility of the systems used for modelling the different layers. Some of the problems and specificities of this area can be viewed as representative for other interdisciplinary fields and are of special interest to us. In particular, the problems related to the integration are discussed in a dedicated section.

3.1.1.11 Complexity-related Issues

Complexity is a factor that is omnipresent. It is neither needed nor not pursued, but nevertheless always present: we can encounter it in every subject domain, in every aspect and on every level of detail. Globalization leads to more competition, therefore more and more features and functionality are built into the products, more and more product variants (especially of compound products) and product ranges are introduced, and the process diversity increases, too. Therefore, complexity arises even where it has not been observed yet as well as increases where it has been present.

Important consequences of the complexity can be grouped as having impact on development, quality, finances, psychology, maintenance, use and others. An attempt to represent them visually is given in Figure 3.4.

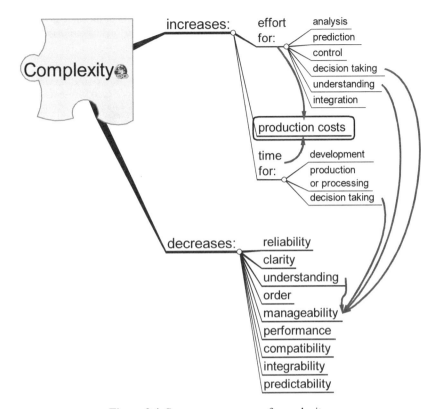

Figure 3.4. Some consequences of complexity

Due to their huge influence on almost every modelling aspect, the complexity issues are discussed in a dedicated section below.

3.1.1.12 Standardization and Standardization-related Issues

3.1.1.12.1 Definitions

3.1.1.12.1.1 Standardization
The standardization process aims to achieve at least the following improvements:

18. Better interchangeability of products, software, processes, documents, *etc.*

19. Better quality, reliability and acceptability of products, processes, services and other goods.

20. Better predictability: if something is said to comply with a certain standard, those who know the respective standard can better tell what is to be expected from it.

Yet, there are some contradictory issues related to standards and standardization. The point is that in order to be efficient and to reach its goals, a standard has to be respected and followed by as many users as possible. In this sense, the so-called *de facto standards* or *industry standards* have a slight lead over the standards designed from scratch. Probably this is one of the explanations of John Sowa's *Law of Standards*, given in Sowa (2004):

> *Whenever a major organization develops a new system as an official standard for X, the primary result is the widespread adoption of some simpler system as a de facto standard for X.*

On the other hand, the needs and requirements of different (potential) users of a standard are usually so different that it is hardly possible to find even two of them that would be equivalent. In order to analyse this situation, suppose we are given a number of enterprises working in the same subject area. Let us consider how we could make them use a given standard in order to achieve the above-mentioned aims.

Assume that a standard is defined through a set of characteristics S_{std} and that each of the N enterprises would have a certain set of characteristics C_E that it desires to have in the standard. Thus we have N sets of characteristics and we want to combine them in a way leading to optimal results. From the point of view of combinatorics there exist three possibilities of how to choose the set which defines the needed standard:

i)   take the union of all N sets $_uS_{std}$:

$$_uS_{std} = \bigcup_{i=1}^{N} C_{E_i} \tag{3.1}$$

j)   take the intersection of all N sets $_iS_{std}$:

$$_iS_{std} = \bigcap_{i=1}^{N} C_{E_i} \tag{3.2}$$

k)   take superset of case 0 which is subset of i) $_iS_{std} \subset {}_mS_{std} \subset {}_uS_{std}$;
l)   appoint a committee, which analyses the subject area, consider the elements of set i), throw away unneeded, controversial or problematic elements and insert additional elements, if considered necessary.

Case i) would lead to maximal theoretical interchangeability, but also to maximal complexity (*cf.* Section 3.1.2 above) and, respectively, to maximal effort for the development of the standard and its support.

Case j) can lead – depending on the specific situation – to an empty set or to a set which is too small to be able to define a useful standard.

Case k) seems to be better as a compromise between the first two cases, but it leads to the problem how to decide which elements of which set are to be taken over and which elements are to be ignored. Some possibilities are discussed below.

And, finally case l) has much in common with case k), but reveals in addition problems of different nature, namely, how to choose the members of the committee, so that all enterprises get balanced representation and their interests can be fairly pursued.

3.1.1.12.1.2 Implementation

Fast implementation of any standard is crucial for its success. Long (implementation) delays can shake the trust in it or cause temporary solutions oust it for many of its potential users. But the more complex a given standard is, the longer takes its adoption and, respectively, implementation. This seems to be the case with, *e.g.*, ISO 10303 (well-known as STEP).

3.1.1.12.1.3 Interface

Each standard ages and at some point in time, a need to extend or update it arises. If this need arises gradually and not simultaneously at all concerned enterprises, some of the affected developers or users of the standard attempt to introduce extensions in their implementation of the standard. Thus, different extended implementations begin to lose their conformity (see the respective section below). Moreover, with each extension from a different implementer, the standard loses its role and strength. One of the consequences is that the interfaces to and from such a standard become inefficient or even non-operational.

3.1.1.12.2 Standard-related Traits

In this section we shall try to define some traits that allow us to assess and compare different implementations of a given standard. In order to provide a better basis for comparison, it is preferable for us to define and use traits that can be measured over the same range – ideally the range [0,1]. Such values can also be interpreted as percentages.

3.1.1.12.2.1 Conformity of a Standard's Implementation

Let there be a standard defined through a certain set of characteristics S_{std} and an implementation of the standard, defined through the set of characteristics *Simpl* which is a union of the set of implemented elements of the standard $S_{sdtElements}$ and the set of implementation-specific elements S_{spec}. As a general trait allowing us to judge how standard-conforming a given implementation is, we can now use the ratio of the number of implemented elements of the standard and the number of all elements in the standard:

$$S_{conf} = \frac{|S_{stdElements}|}{|S_{std}|} \tag{3.3}$$

Here S_{conf} is the standard conformity, which has a range [0,1].

3.1.1.12.2.2 Specificity of a Standard's Implementation

This trait is not just the inverse value of standard conformity, but also a measure of uniqueness. Therefore, it is proportional to the number of implemented extra characteristics (extensions of the standard) and to the number of unimplemented characteristics, but is inversely proportional to the number of implemented characteristics.

$$S_{implSpec} = f\left(\frac{|S_{spec}|}{|S_{std}|}, \frac{1}{|S_{impl}|}\right) \tag{3.4}$$

3.1.1.12.2.3 Interfaceability of a Standard

It could be interesting to estimate how difficult it is to interface two different implementations of the same standard, *i.e.* to establish a way of exchange between them. This trait, being specific to each set of a given standard and any two of its implementations, will be called interfaceability and will be denoted *I(standard,source,target)*. It is proportional to the standard conformities and inversely proportional to the specificities of the both implementations:

$$I_{(std,S,T)} = f\left(SC_S, SC_T, Sp_S^{-1}, Sp_T^{-1}\right)$$ (3.5)

3.1.1.12.2.4 Standard-related Complexity

No two standards have equal complexity – each standard is specific and has its peculiarities. Yet, efforts to reduce their complexity are observed. The introduction of application protocols in STEP, for instance, aims at reducing the complexity by restricting the scope of each protocol and thus making its implementation easier, inasmuch as several granules of smaller size are to be considered instead of a huge one.

3.1.1.12.2.5 Effort for Exchange/Porting

The standards for data exchange are being designed and developed to facilitate the exchange and make it more efficient. With increased standardization effort, though, the total effort for exchange and standardization may become bigger than the effort for exchange without standard. Based on the author's observations, in Figure 3.5 a hypothesis is presented about the total effort with regard to different standardization grades. According to this hypothesis the minimal total effort for exchange is expected to be somewhere between conventions and meta-norms. The exact position depends on the specific case, scope as well as other factors.

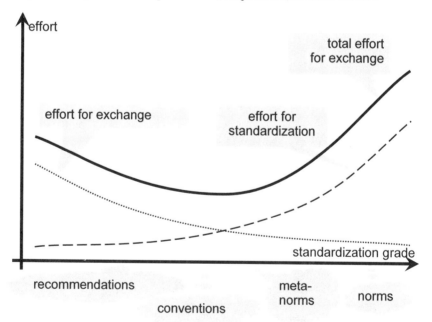

Figure 3.5. Effort for exchange/porting depending on standardization grade

3.1.1.12.3 Compatibility and Data Exchange

The development of a good standard takes time – everything should be carefully contemplated, verified, validated, *etc.* This time can be so long that the requirements can change or even a need for a totally new standard could arise. A typical example here is ISO 10303, whose first version development time took more than a decade, and even before it was complete, the work on the second version started. Despite great achievements of ISO 10303, the main problem nowadays is still the incompatibility of the data formats of different CAx-systems. On the one hand, it is attempted to overcome the incompatibilities by development and use of standards for data representation and exchange. On the other hand, in order to integrate models created in different systems, an exchange between the systems still has to be performed. An exchange using a standard format would just save the converter (*cf.* Figure 3.6) but not the exchange itself. This means that the mentioned architecture would still have problems with the huge size of the models.

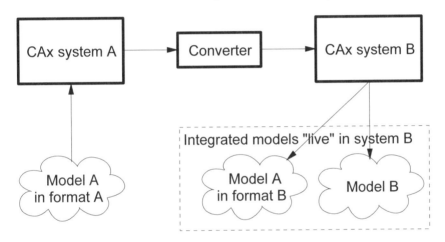

Figure 3.6. Integration of models by means of data exchange

Note the paradox: STEP's objective is "to provide a (*system-independent*, or *neutral*) mechanism that is capable of describing product data throughout the lifecycle of a product, independent from any particular system", but no CAx-system producer is even thinking about changing the native format of its CAx-system to STEP, let alone doing it! This means that either the so-called STEP pre-processors and postprocessors or the PDM-enablers (or both) have to exist for each CAx-system type.

Another point, which is often disregarded, is discussed in Avgoustinov (1997). It is never the case that all data in a production chain has to be available (and respectively – converted) to all CAx-systems of the chain. Moreover, due to the specifics of the information flow it might be more efficient to have distinct standards for the exchange (*cf.* Figure 3.7) *within* each phase and perhaps a few standards for inter-phase exchange, rather than having one standard to cover the whole development and production chain.

Any standard acts as an accelerator up to a given moment in the development, but after some (the same or other) point in the time it acts conservatively and slows down the development unnecessary, especially by wrong subject or extent. In many cases a set of carefully chosen conventions can suffice and be much more flexible than a standard. On pp.166–167 of Booch *et al.* (1999), for instance, it is suggested that "A well structured interface is *simple yet complete,* providing all the operations *necessary yet sufficient* to specify a single service;…" (parts of the citation italicized by the current author). And the increasing popularity and success of industry standards like **CORBA**[23] (*cf.* **OMG (1998))**, **DCOM+**[24] and **Enterprise Java Beans**[TM25] confirms again the flexibility and the acceptance of these alternatives to a "global standardization".

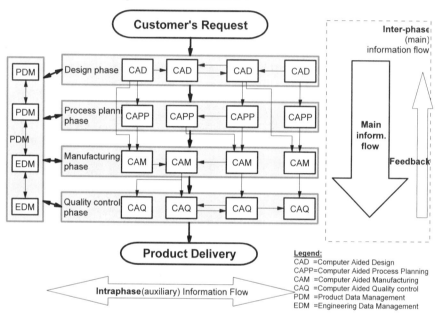

Figure 3.7. Standardization of the information exchange within a production chain: dividing in several sub-domains, after Avgoustinov (1997)

3.1.1.13 Other Problems

There are problems related to security, complexity, diversification, heterogeneity, usability and overall value. Many of them are consequences of globalization, which intensifies the competition by increasing the pressure of costs, the pressure of time, the need for cooperative work and, therefore, the need for communication. Moreover, globalization exposes some specific problems like norm and regulation diversity among the partners, clash of cultures, as well as some special requirements. Also not to be ignored is the group of *psychological problems*: using

---

[23] Abbreviated from Common Object Request Broker Architecture.
[24] Abbreviated from Distributed Common Object Model. It is Microsoft's architecture for working with distributed objects.
[25] The name of the architecture for working with distributed objects from Sun Microsystems, Inc.

software can be fun, but when the software is unexciting, complex, buggy and lacks ergonomics, its use can be boring. We should not forget that many new technologies, products or ideas fail to become popular due to secondary factors like lack of attractiveness, (critically) high accidental complexity or bad advertisement and popularization. Quite to the point, Brooks Jr. (1987) spoke about the role of exciting technologies and their "great designers". Considering, on the one hand, the high number of industry standards created by a couple of people but having great success, and on the other hand, the large number of unsuccessful standards created artificially by committees, he argues that "we should better learn to *grow great designers*" than to develop "great" standards and tools (quotation complemented by the author).

3.1.2 Complexity-related Issues

> *The challenge over the next 20 years will not be speed or cost or performance; it will be a question of complexity.*
>
> Bill Raduchel, ~1999
> (Chief Strategy Officer, Sun Microsystems)

> *Our enemy is complexity, and it's our goal to kill it.*
>
> Jan Baan, ~1999
> (SAP competitor)

> *Simplicity is the ultimate sophistication.*
>
> Leonardo da Vinci, 1452 – 1519

The fact that complexity and complexity-related problems have been in the focus of many investigations from ancient times until today illustrates the importance of this topic. There exist different views on complexity, methods to oppose it and even sayings about it. One aphorism says that most genius-made things are simple, although not all simple things are works of a genius.

According to Brooks Jr. (1987) "*...from the complexity comes the difficulty of communication among team members, which leads to product flaws, cost overruns, schedule delays... less understanding ... unreliability*". As complexity leads to many other problems and inconveniences (*cf.* Figure 3.4 above), it is very important to try to analyse the influence of complexity on the modelling and develop methods to measure and reduce complexity. But what is complexity?

3.1.2.1 Towards a Definition

Although a common word such as complexity should be intuitive enough, a brief look in the relevant literature shows a wide spectrum of interpretations of this term, varying from "the opposite of simplicity" to long and domain-dependent definitions. Furthermore, complexity is discussed in different context by different authors: *e.g.*, in Fagade *et al.* (1998) we find "total product (or project) complexity", "management complexity", "design and manufacturing complexities" and "process complexity"; or in Edmonds (1999b), offering a study of the complexity definition by various authors, it speaks of "true complexity", "real complexity", "system complexity", "observer complexity", "computational complexity", "Kolmogorov complexity", "descriptive complexity", "complexity *per se*"; Grünwald and Vitányi (2003) use, in addition, "descriptional complexity"[26], "object's complexity", "algorithmic complexity" and "stochastic complexity"; finally, we could supplement all these with essential and accidental complexity after Brooks Jr. (1987) (*cf.* previous paragraph and Section 3.1.1.2 above). To summarize, there is no unified view on complexity: scientists could be distributed into at least three groups according to their understanding of complexity.

The views of one group (mainly IT-experts) would be, probably, best illustrated with the definition, given in Black (2005): "*The intrinsic minimum amount of resources, for instance, memory, time, messages, etc., needed to solve a problem or execute an algorithm.*" This definition is not tangible enough to use as a base for (more) quantitative assessments. And since it measures the complexity of a problem only indirectly, one could get the impression that the complexity depends on the tool used for its solution, which is unacceptable. Another definition of complexity in the same sense, but explicitly bound to the (context of) algorithms is given in Howe (2006): "*The level in difficulty in solving mathematically posed problems as measured by the time, number of steps or arithmetic operations, or memory space required (called time complexity, computational complexity, and space complexity, respectively).*" As we can see, both definitions mainly refer to *computational complexity*, which constitutes only one part of the complexity of a software model.

Another widespread interpretation is that complexity can be viewed as a *measure of the difficulty to understand a given matter or to deal with it.* The lack of understanding alone is not always a problem, but it becomes important as soon as a decision based on the respective matter has to be taken. Since understanding can hardly be quantified and, in addition, depends on other factors (*cf.* Figure 3.14), such a definition is of little use.

A better alternative is proposed by another group, represented, *e.g.* by Suh (*cf.* Suh (2001), chapter 9). According to this definition, complexity and information are tightly related and:

Definition 3.1:   *Complexity is a* measure of uncertainty *in achieving the desired functional requirements.*

This definition allows us to measure complexity directly in bits, which is very convenient and puts the stress once again on the relation to information. Thus, it is applicable not only to software systems, but to any other system with a countable number of components. However, it is measured or defined as "*only relative to what we are trying to achieve and/or want to know*" (*cf.* Suh (2001, p. 472), meaning that until the functional requirements are known the complexity either

---

[26] Cited as used in the source; not to be confused with "descriptive complexity"!

does not exist, which is obviously false, or we cannot measure it, which is rather inappropriate. Yet, complexity according to Definition 3.1 refers only to a part of the whole complexity phenomenon and should probably be better called "complexity of achieving the aims".

Even the simplest technical system or the simplest object can be "finite" only at a certain level of observation or abstraction. For instance, we are not able to count the atoms in the smallest mechanical part or in the smallest piece of matter. Physicists have not succeeded until now to fully explain the structure of the atom itself. It is (still) impossible to repudiate certain "fractality" of the matter – *cf.* the Bohr model, explained in Wiki (2006) as a planetary model in which the electrons orbit a tiny nucleus in the way that the planets orbit the sun. Indeed, neither in the direction of the macrocosm nor in the direction of the microcosm an End comes into sight, meaning that – as for now – we can hardly dream of final (or countable) systems. Consequently, the real complexity of anything tends towards *infinity*, or at least is immeasurable. So, Edmonds (1999b) states: "*The "true complexity" of real objects (if it existed) would probably be totally beyond us*".

Nevertheless, in our everyday life we perceive certain things as more complex than others, or some aspects of the same thing as more complex than others – *e.g.*, the use of any product is normally much easier than its development or production. Moreover, we can deal (although not in all cases) with this "infinite complexity". What helps us is our ability to abstract away from the inessential details and deal with the essentials only. For this reason I shall introduce a new term here:

Definition 3.2:   the perceivable or ascertainable part of the absolute (or full, or total, or overall) complexity of an entity will be called discernible complexity *of this entity.*

The term *entity* is used as a generic term (or placeholder) because it is more general than model, product, process, *etc.* The entity could also be a task – *e.g.*, to prepare a model or to produce certain product. Therefore, all references made to an entity within this section could be applied as well to any problem, task, model, product, process or other complex object.

As already mentioned above, many different complexity types are discussed in the literature, but no attempt of their systematization or their ordering in a taxonomy has been available until now – *cf.* also Weber (2005c). In my view, when speaking about one and the same entity, there are no different types of complexity, but different views or aspects on it.

Definition 3.3:   any part of the discernible complexity of a given entity, which is important or essential, mainly in connection with something specific or from a specific viewpoint (aspect) will be called aspect complexity.

Development, use, testing, marketing, *etc.* are examples of different aspects. Each aspect appears to be related to (specific) activities, and activities are often specific to a given phase within the lifecycle. The overall discernible complexity, which is relevant for a given problem or entity, can thus be viewed as greater than or equal to the sum of the complexities of all its aspects. In addition to the discipline-specific aspects there could exist general or product-specific aspects, too.

In interdisciplinary branches of science like mechatronics, one or more of the general aspects may exist in more than one of the involved disciplines, which additionally increases the complexity. In this text, "discipline" is used as a more general term than "domain". To avoid possible confusion, all (or most) discipline-specific parts of such interdisciplinary models are separated in their own *layer* (*cf.* the definition of layer in chapter "Modelling basics").

We could think of aspects and disciplines as of two dimensions forming the *complexity space*[27] of a given entity, as illustrated in Figure 3.8. In fact, there is one more relevant dimension of complexity that is not considered here, namely, structure (to be discussed in Section 3.1.2.2.3 below), but it is difficult to prepare an adequate representation on (two-dimensional) paper.

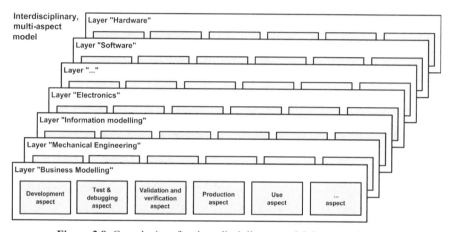

Figure 3.8. Complexity of an interdisciplinary model: layers and aspects

Until now, the causes of complexity and the factors influencing it have been investigated unexpectedly little. So, let us start with some observations.

3.1.2.2 Observations

When we use the word "complexity" alone we actually (unconsciously?) mean "discernible complexity". The absolute complexity of software models is always finite (after all, they are created on computers having final memory and saved on a medium within finite place), but often it is higher than we can perceive, feel comfortable with or would like to have.

The brain cannot perceive information having (discernible) complexity above certain *critical level*. This critical level is person-dependent, *i.e.* the discernible complexity is subjective. In particular, it depends on the person's *a priori* knowledge (depending in turn on education and background), on the power of comprehension, and perhaps – at least to some degree – on the training. An illustration of the different perception of two persons, experts in the same competence domain, is given in Figure 3.9.

---

[27] Not to be confused with the term *space complexity*, used in computer science to refer to the amount of memory space that a computer program requires for its proper execution.

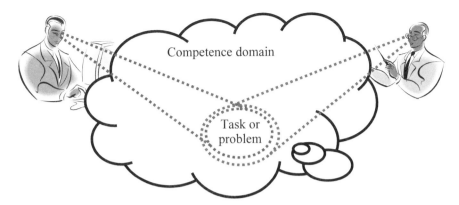

Figure 3.9. Individual limits by perception of complex matter

In order to understand why the critical level of complexity is essential, suppose that we are observing (only) one person, having to solve a problem or a task in the considered competence domain – say, to develop a product for us. We call this person (at least for this example) Solver; and we – as a customer – are permanently increasing the number of our requirements for the negotiated product.

As expected, with the increased number of requirements the complexity of the problem increases, too. At the beginning the problem is simple (its complexity is negligible), and the problem is solved in the best possible way. With increasing complexity, we reach a point when our Solver starts to produce sub-optimal solutions (or products). This means that he either cannot find an optimal solution, or finds multiple solutions but cannot decide which is the optimal one. If we continue to pose new requirements (and thus further increase the complexity), a moment will come when the complexity is so high that the Solver cannot control it anymore, and either delivers a wrong solution or cannot deliver a solution at all. This stage corresponds to the critical complexity level.

The ability to make (appropriate) decisions as a function of the complexity is graphically represented in Figure 3.10.

The critical complexity level marks the boundary between solvable and unsolvable problems, or between ability and inability of making a decision. So it makes sense to terminologically distinguish whether the complexity is below or above this level. Since the ability to take decisions is typically related to the ability to control a given situation, we shall use the terms controllable and uncontrollable for complexities below and above the critical level, respectively.

In general, a typical sign that the critical level is reached is the fact that the person in question either ceases to perceive (the excess of) information or cannot start to perceive it at all. The latter case happens typically when the information comes from a foreign domain about which very little *a priori* knowledge is available. The additional uncertainty, resulting from the lack of knowledge and understanding that are needed to solve a given problem, is called *imaginary complexity* in Suh (2001, p.476).

The imaginary complexity is usually higher when multiple aspects of the modellee have to be considered. If the discernible complexity of the main aspect of the modellee is already near to the critical complexity level of the Solver, this means that every new aspect that has to be considered increases the risk of reaching a situation without (optimal) solution. How to deal with this challenge? If we assume that the critical complexity can be expressed as the number of facts that are (still) comprehensible at the same moment, one possibility to reduce or keep the complexity below the critical level is visualized in Figure 3.11.

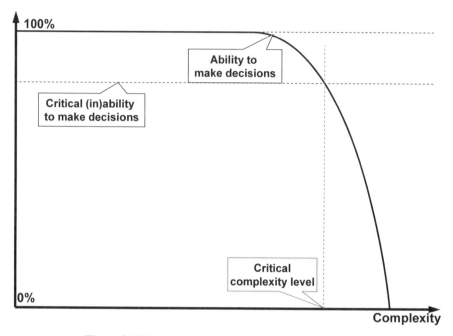

Figure 3.10. Dependence of decision making on complexity

The complexity of a process and its results are not necessarily mutually dependent. Often simple procedures (codes, programs), both in nature and computing, yield most complex results, and *vice versa –* complex procedures return simple results. Similar observation can be made about the interdependence of the complexity of a problem and its solution. Therefore, a reasonable question is what exactly generates and influences the complexity of a model.

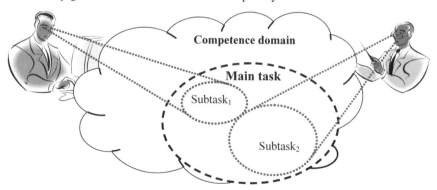

Figure 3.11. Focusing by splitting/reducing the scope to keep the complexity controllable

3.1.2.2.1   Factors Influencing Complexity

This section is an attempt to show at least some of the factors directly influencing the complexity (of any model).

Edmonds (1999a) enumerates thirteen "Unsatisfactory Accounts of Complexity" (pp. 57–67). Many of them are, in my view, either just symptoms of existing complexity (*e.g.*, "irreducibility" or "ability to surprise") or consequences thereof (*e.g.*, "improbability", "processing time", "ignorance").

The discernible complexity of an entity (modellee, model, *etc.*) depends on the *competence domain* of the respective viewer (*cf.* again Figure 3.9). The factors can be divided with regard to their dependence on the modelling domain into two large groups: domain-independent and domain-dependent factors.

An example illustrating the distinction in perceiving the complexity based on a competence domain is the key difference between the creator and the user of the same entity. In general, each of them has different background and purpose, and perceives completely different complexity of the (same!) entity (*cf.* Table 3.5 below). Here, it certainly makes sense to speak about role-dependent complexity.

3.1.2.2.1.1   Domain-independent Factors

The factor with the strongest impact on the complexity of a model (*cf.* also Figure 3.14) is the discernible *complexity of the modellee*. In turn, it is a function of the number of (discernible) components in the modellee and of the number of (discernible) attributes (or features[28]) in each atomic component. So, let an attribute be defined as follows:

Definition 3.4:   *a quality or property of an atomic entity (like model, modellee, component, etc.) that is distinguishable from the rest of the entity is called* attribute.

For instance, any line segment drawn on paper is atomic and has two discernible ends which, consequently, are (the main) attributes of the segment. A software model of the same line, though, is not atomic. It contains at least the mentioned attributes (the line ends) represented by means of a couple of variables or by means of models of other geometric elements – *e.g.*, models of points.

An attribute need not be visible in order to be discernible – *e.g.*, a circle's centre can be invisible (or not visualized), but we know that it exists and it can always be determined if more than two points on the circle are known. Consequently, it is also discernible.

Processes have attributes too. For instance, the moments of start and end of every process are well distinguishable in time and, therefore, are also attributes.

Another example, illustrating the direct dependence of the discernible complexity on the number of (modelled) components and attributes, is a system of arbitrary number of points and line segments that can connect any pair of points. It is clear that by increasing the number of points the complexity of this system will also increase. This increase can be perceived also visually in Figure 3.12 (from left to right).

---

[28] The word *feature* is probably a more adequate term for what I mean here, but it is too overloaded with other meanings and would lead to misunderstandings. For this reason I prefer to use the term *attribute* instead.

Figure 3.12. Systems of points and connecting lines with increasing (from left to right) complexity

Note that the complexity of a system depends on the number of components – or attributes (in the above case, points) – not linearly. This might be not discernable at first glance, but it becomes apparent if we consider that the second system in Figure 3.12 can have two more line segments – corresponding to both diagonals – while still having only the same four points. Besides, any of the four line segments shown in the figure may be omitted, as illustrated in Figure 3.13.

Figure 3.13. Systems of 4 points and different combinations of connecting line segments

Let us have a closer look at how the two rightmost systems in Figure 3.13 are composed. Two of the line segments have been crossed and create thus a new attribute or component – their crossing point – which additionally increases the complexity of the affected system. We could certainly pretend that this (virtual) point does not exist, exactly as we could pretend that the centre point of a circle does not exist, but this does not reduce the complexity. I shall call such unpredictable complexity a *derived complexity*. It is part of either the essential or the accidental complexity (or both), which are, in turn, parts of the discernible complexity. In general, in a system of points with some points being connected with lines, it is not trivial to predict whether any pair of lines will get crossed or not. On the contrary, given a system of more than 3 points where a) all points lie on the same plane, b) no three points lie on the same line and c) every two points are connected with lines, it is certain that there will be at least one pair of lines creating new point(s) by means of crossing.

Suppose that the points represent notions (or concepts, or models, or whatever) and the connecting line segments represent the (possible) relations among these notions. It is clear that – desired or not – similar phenomena can lead to derived complexity, meaning that the overall complexity of the system also increases. Apparently, by such a detailed modelling, a more complex modellee would get a more complex model. And only rarely, we can reduce the complexity of a model by means of reducing the complexity of the modellee.

The factor with the second-strongest impact on complexity is probably the *purpose of the model*. Although it influences the complexity only indirectly, the purpose of a model determines the *requirements* for it and the necessary *level of (modelling) detail*[29] – *i.e.* what must be modelled and what could be neglected. If

---

[29] Not to be confused with level of (visualization) detail, as it is used, *e.g.*, in the Virtual Reality Modelling Language (VRML). The latter allows us to achieve better visualization

we model something in order to understand it, we do not need many details. If the purpose of the modelling is to optimize the modellee, more details are needed and have to be modelled, and the complexity of the resulting model grows.

Let us consider the representation of a circle on different kinds of (computer) devices. From a mathematical point of view, any circle is defined as all points having the same distance – the radius – to a specific point – the centre. Now, the infinite number of points comprising any circle are to be represented in a time that is not only finite, but also acceptably short. Without use of compasses, this can only be achieved as a compromise with the representation accuracy. Therefore, at least two different circle models are typically used. The one is used for saving on a medium and is similar to that in Figure 2.34; being compact and unambiguous, it is used for *internal representation*. The other is used for visualization purposes or *external representation* on output devices and is normally derived (or automatically generated) from the internal representation. For the visualization, any circle is typically modelled by a polygon, where the number of points can vary in certain limits and is by and large a variable parameter for achieving flexibility. More points mean more accuracy, but also more calculations. Less points mean less calculations, but an inadequate choice of this number – *e.g.*, less than 8 points – leads to visualization of other well-known geometrical figures as in Figure 3.12. Thus, a quantity reduction can lead to quality reduction or even to loss of information.

Immediately bound to the requirements and to the level of detail is the *number of modelled functions and properties* of the modellee. The developer of a model could vary this number and choose what exactly to model (or implement) – but only to a limited extent in order to fulfil the purpose of the model. If we again consider the circle model in Figure 2.34, we note that no appearance properties (colour, line type, filling, centre marker, *etc.*) are modelled. The required memory and the complexity are kept low in this way. But the functionality, the accuracy and – to some extent – the adequacy of the model are reduced.

In turn, the number of modelled properties determines the number of model variables and their type. We can assume with acceptable accuracy that each important attribute of the modellee will be represented in a software model by means of one variable, but this variable has to be of appropriate type. If we consider the model of a circle from the previous chapter (*cf.* Figure 2.34) again, we note that the radius is represented by a numeric (floating point) variable, while the name is represented by a text variable. The centre of the circle is represented by a variable of type point, which is actually a compound variable built up from two numeric variables encoding the two coordinates of the centre in a Descartes coordinate system. Compound properties are represented by compound variables, so that each attribute has a corresponding variable, while there could be variables, having no corresponding attribute and being used for internal purposes of the model only. Nevertheless, the (data) types of the variables have impact on the complexity too. An overview of different concepts and term, related to complexity, as well as their interrelations are presented in Figure 3.14.

---

speed by choosing the most appropriate level of modelling detail (out of several different levels), when it is not necessary to show the model in the full possible detail. Since this technique introduces redundancy (part of the information is repeated in each encoded level), it actually increases the model complexity.

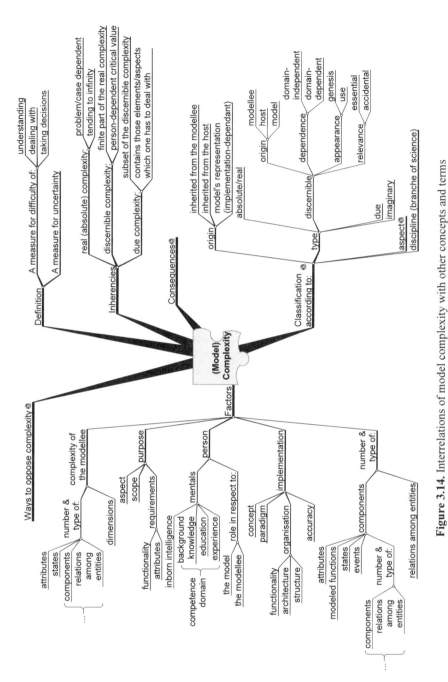

Figure 3.14. Interrelations of model complexity with other concepts and terms

Similar considerations can certainly be made with regard to the functions that have to be modelled. Simply any function (or activity, or behavioural element) has to be represented by a program, because a single (compound) variable is inadequate and insufficient to represent even the simplest function.

3.1.2.2.1.2 Domain-dependent Factors

Another factor of influence is the implementation of the model, which is always domain-dependant. Since the developer is usually free to choose arbitrary implementation (he must only deliver the required functionality and features), a proper choice could play a significant role in reducing complexity. Actually, the implementation has almost no influence on the essential complexity but only on the accidental one (*cf.* Section 3.1.1.2 above). Thus, the choice of an implementation strategy has to be a well-considered compromise between options (of approach and tools) and their complexity.

Let us consider some specificities of the software models and their implementation. An analysis of the software models and their origins reveals a domain-specific stratification of the model representation. The following levels can be recognized:

m) Hardware
n) Machine language, running (and depending) on m)
o) (Macro) Assembler, extending n)
p) High-level language, written in o)
q) Application written in p)
r) Application-based model living in q)

What all these layers have in common is that all they are involved directly or indirectly in the representation of a software model. And no matter how complex the model is, it is represented on the lowest level through ones and zeros. As a rule, *unnecessary complexity should not be tolerated on any level*. If the layers q) and r) are more open or if the modelling takes place in layer p), it is possible to achieve the desired flexibility and robustness to a much higher grade.

In the case of CAx-applications – and CAx-systems are applications too – the closest level would be q), followed by r). On the other levels it is possible either to see the source code or to debug, or both. On level q), however, this is impossible, and on level r) this is only partially possible, and to what extent it appears to be case dependant. Such closeness on the top of the hierarchy has strong impact on the flexibility and also on all data exchange or data integration issues.

Let us consider again the lifetime of the elements of the information processing, presented in Section 3.1.1.8 above. Due to apparent correspondence to the levels m) to r), it is possible to view these elements as sub-domains of the software-modelling domain. Since the models are to be viewed as data or data-derivatives in this hierarchy, it is clear that their representation depends on (the lifetime of) all lower levels. In this situation long lifetime of the model representation can be achieved only if there is succession from one generation to the next within any lower level. Such succession is often called backward compatibility.

3.1.2.2.2 Components of Complexity

At this stage it seems that the absolute (or full) complexity of anything consists of at least three components, or three kinds of complexity:

s) *Imaginary*; may also be called *spurious*, because in many cases it can be reduced to zero by means of learning.
t) *Random*: unpredictable fluctuations (of system parameters) increase the total uncertainty.
u) *Combinatorial*: depends on a finite number of factors whose possible permutations also increase the total uncertainty; alternative name is *computable*, since usually it can be computed.

Each of these components increases the overall uncertainty in dealing with the respective matter.

3.1.2.2.3   Distribution (Analysis) of Complexity

In order to process an entity of critical complexity, one usually splits it, for simplification, into several sub-entities (*i.e.* applying the well-known "divide and conquer" principle). After the split, each of the created sub-entities has lower complexity and their sum equals the complexity of the initial entity. It is possible to repeat the same operation over and over agan, and thus to build hierarchies of entities.

Suppose the complexity of the main entity on the top level is one (100%). If we split this entity into (two) sub-entities (SE_1 and SE_2 in Figure 3.15). The sub-entities need not be equal in complexity, but it would be advantageous, as we shall see later. Yet, the sum of their complexities will remain equal to the overall complexity of the parent entity.

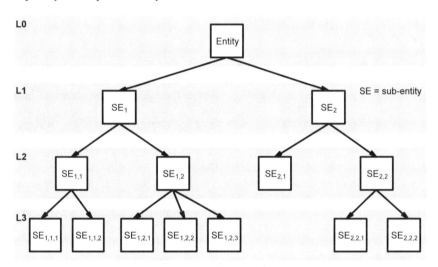

Figure 3.15. Distribution of the complexity within a given hierarchy

A level complexity is the sum of the complexities of all sub-entities within a given level and is equal to the initial complexity of the top entity (each level has yellow background in Figure 3.15 and an index on the left). Since lower levels have larger number of sub-entities, the average complexity there is also lower. The complexity on the lowest level (*i.e.* in each leaf of the tree) should be lower than the critical complexity for the person that should process the entity. Thus, the complexity of a sub-entity in a similar hierarchy will depend on its level within the hierarchy and on the "regularity" of the complexity distribution within each level. It is worth noting that in contrast to the trees of data structures, a hierarchy-tree of an entity or a model cannot have nodes with only one child. This results from the nature of the process – we are splitting entities for splitting their complexity, but nothing should get lost, therefore, a "split into one part" is no change.

Hierarchies, where the complexity of the separate sub-entities within any level is approximately the same and the number of sub-levels in the branches of each node is equal, are *well-balanced* and easier to handle – *cf.* Figure 3.16.

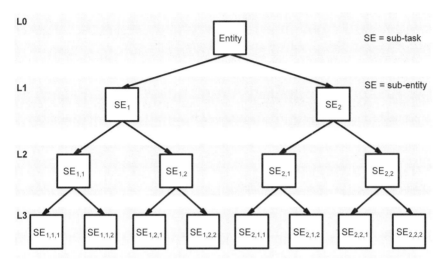

Figure 3.16. Balanced distribution of the complexity within a given hierarchy

For both cases illustrated in Figure 3.15 and Figure 3.16 we can write:

$$complexity_{L0} = complexity_{L1} = complexity_{L2} = complexity_{L4} \qquad (3.6)$$

In addition, for well-balanced trees it is easier to calculate the (average) complexity of the leaf-nodes (those that have no sub-nodes): it is simply $1/LN$ of the complexity of the root node, where LN is the number of leaf-nodes. When the entities are always split into *two* sub-entities, the tree becomes a binary-tree and has all specificities of the binary-trees. Thus, we can calculate the number of leaf-nodes LN for a binary tree with L levels as:

$$LN = 2^{(L-1)} \qquad (3.7)$$

If we index the levels from the root down and start from zero as in Figure 3.16 (note that levels are labelled with $L0$, $L1$, $L2$ and $L3$, but the indices are 0, 1, 2 and 3, respectively), we can use the index i of any level for calculating the number of nodes N_i in it:

$$LN_i = 2^i \qquad (3.8)$$

Furthermore, it is possible to connect the number of levels L, the number of leaf-nodes LN, the average complexity of leaf nodes C_{avg} and the complexity of the root node C with the following formula:

$$C = C_{avg} * LN \qquad (3.9)$$

or, alternativelly

$$C = C_{avg} * 2^{(L-1)} \qquad\qquad (3.10)$$

The Equation 3.9 can also be used for estimation of the complexity of cases, where the structure is not a balanced tree, or even not tree-like at all. But generally it is useful to put $C=1$ (or 100%) and look at the graphical representation of the interdependency of C_{avg} and LN, given in Figure 3.17.

As we can see on the graphic in Figure 3.17, dividing an entity into sub-entities leads to decrease of the average complexity, but the dependency is not linear. Initially, it is very efficient; after reaching a number 10 to 15, the fall of the curve decreases rapidly, and for more than 20 leaf-nodes the fall of the average component complexity is so slow that further fission is hardly worth the effort.

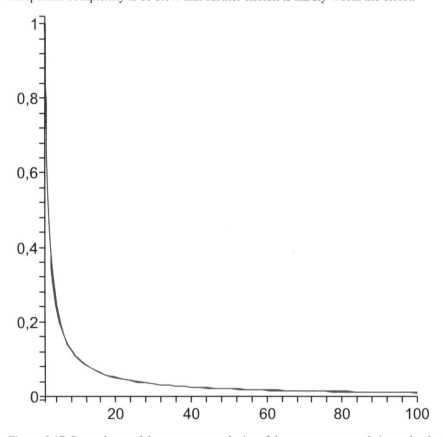

Figure 3.17. Dependence of the average complexity of the components on their number by keeping the overall complexity constant (100%)

Thus, the distribution (splitting) of complexity among many sub-components leads to simplification of the (resulting) components. More simplicity leads to easier analysis, which, in turn, leads to better understanding and finally to better (modelling) results.

3.1.2.2.4 Decomposition as a Way for Problem Solving

For the above-mentioned reasons the method of decomposition (or divide and conquer strategy) is very popular for problem and task solving. The process can be illustrated as in Figure 3.18

A problem is decomposed into sub-problems until either there is a solution available for every sub-problem or a (sub-)solution can be found for a given sub-problem. These two processes constitute the *analysis phase* of problem solving, but as can be seen in Figure 3.18, this is only halfway to a full solution. After a solution for each sub-problem is available, the sub-solutions have to be put together or integrated. Since this process – similarly to solving a puzzle– involves a great deal of combinations and matching the sub-solutions to one another, it is denoted in the figure as composing/combining. At the end of the process we have a solution – represented by the assembled puzzle-parts at the lower left corner in Figure 3.18. The circle around the puzzle-parts symbolizes their unity. Now, as this solution has to be validated or at least compared with the problem, we have again matching. In reality, a solution seldom matches the problem exactly, which is symbolized by the different contours of the respective graphical representations in Figure 3.18. Thus, the composition and solution-problem matching constitute the second phase of problem solving – the *synthesis phase*. The problem solving can be viewed as a cycle too: when the problem and its solution are too different, the difference can be seen as a new – usually smaller than the original – problem to be solved.

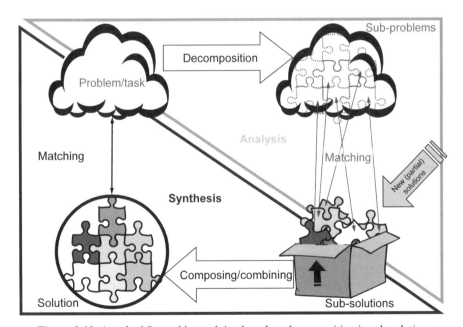

Figure 3.18. Amethod for problem solving based on decomposition in sub-solutions

Another symbolism in Figure 3.18 is the size of the sub-solutions, making allusion to the granularity. Clearly, a coarse-grained solution would be put together faster but match the problem worse than a fine-grained solution. Yet, there are other specificities of the synthesis which deal with complexity too, and are discussed in the following section.

3.1.2.2.5 Composing (or Building up) Complexity

The distribution of the complexity discussed in the previous section and the divide-and-conquer strategy help us to resolve complex problems or tasks into smaller ones, and thus simplify the steps towards a complete solution or a result. But this (simplification) is just the one side of the coin. The other side is, of course, the reverse process of combining (already) existing models and entities into compound models, entities, *etc*. These kinds of activities, leading to more complexity of the resulting entities, are known as *synthesis* and are typical when (unit) construction sets are used.

In Section 3.1.2.2.2 above we assumed that each entity splitting was ideal in the sense that the complexity was fully distributed among the new parts resulting from the division. However, when we try to combine or integrate entities (*i.e.* in a system), the situation can be slightly different, especially if the entities are designed not for each other but as universal components instead. Even if these entities are just two, they may need an additional entity for holding or putting them together. Let us give an example of a similar situation for comparison and illustration. A material object, *e.g.*, a wood stick, can be easily broken in two pieces. If we want to reconstruct the initial stick from the pieces, we need some kind of glue as an additional entity. Depending on the material of the object, we may use a solder, a brace, a simple container or something else instead of the glue. In software systems, the area of a module that allows us to connect it to another module is called interface, and the role of the glue is played by interface modules.

3.1.2.2.6 Aspects of Complexity

As already mentioned (*cf.* Definition 3.3), the discernible complexity depends not only on the person, but also on the aspect (or context). We can distinguish as many complexity-related aspects as we need or desire, which is illustrated in Figure 3.19.

It seems impossible, though, to quantify the overall (discernable) complexity without quantifying each related aspect first. On the other hand, it is apparent that the complexity of different aspects of the same entity may be different – some examples are given in Table 3.5. In the first column, example entities are listed; in the first row of columns 2, 3 and 4 – the aspects *creation*, *production* and *use*; and in the remaining cells – actions that are specific for the respective example and aspect.

Exactly as mathematical operations on objects of different types are not always meaningful, comparing the complexity of two different objects without specifying the aspect or of two different aspects of the same object is not always meaningful. For this reason, the importance of the ability to assess and quantify different complexity aspects increases. Special attention deserve those aspects that are present in (almost) all types of models.

The discussion of all aspects mentioned in Figure 3.19 goes well beyond the scope of this work, but let us at least try to order them in a simple structure, based on the phases of a model lifecycle. A feasible grouping is sketched in Figure 3.20 with regard to which model lifecycle phase a particular aspect belongs to. Note that the relations between some pairs of aspects are directly indicated here. The numbering of the aspects reflects their approximate succession within a model lifecycle. The global aspect, for instance, factors out and groups those complexity sub-aspects that are valid for almost every lifecycle phase, and it is assigned the number 1 in Figure 3.20, and further aspects are assigned the numbers from 2 to 8. The aspects with numbers between 5 and 8 are not always present, therefore they are grey.

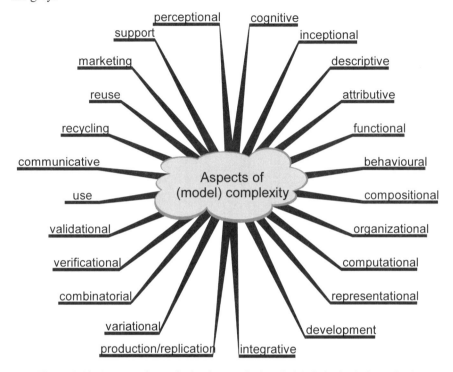

Figure 3.19. Aspects of complexity that can be handled (relatively) independently

Table 3.5. Example (common) aspects of different sample entities

Entity	creating	producing	using	
			Aspect:	
meal	inventing	cooking	degustation	eating
sword	designing	forging	teaching duel	fighting a duel
medicament	developing	producing	prescribing	taking
car	designing	manufacturing	driving	riding
aircraft	designing	manufacturing	piloting	flying
car factory	planning	building, equipping	maintaining	producing cars

For the purposes of this work let us look into those aspects that are more closely related to modelling. The order in which complexity aspects are discussed below differs from that in Figure 3.20, and is derived from the (in)dependence of one aspect from another. Let us consider when, where and how the complexity of a given (sub-)aspect can be computed.

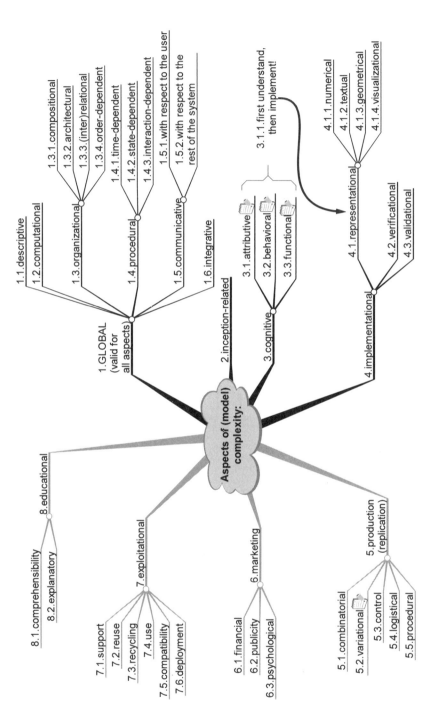

Figure 3.20. A simple taxonomy of aspects of complexity

3.1.2.2.6.1 Inception-related Aspect

Every beginning is difficult, due to various causes – among others also psychological, technical, cognitive, *etc*. Yet, most of them come out of the opposition between two types of categories, namely: *reasons to start* (need, requirements, customer demands, *etc*.) and *prerequisites* (or possibilities to meet the reasons). As soon as each element of the former set is satisfied through an element of the latter set, the formal conditions for inception are fulfilled. The inception-related complexity apparently depends on these two sets, even if this is hard to express mathematically.

3.1.2.2.6.2 Cognitive Aspect

If we assume that complexity aspects arise in the order illustrated in Figure 3.20, it is indeed arguable whether the cognitive aspect or the inception-related aspect comes first. On the one hand, knowledge about the problem domain is a prerequisite for the possibility to elaborate a solution, and in fact the preparation of any model is a kind of solution as well. Yet, in many cases the necessary knowledge (or a large part of it) is acquired only after the start of the work. The latter is especially true when preparing a model of a virtual modellee.

As indicated in item 3.1.1. in Figure 3.20, no implementation may begin before the matter is understood. The process of thinking and cognition is almost impossible to grasp, describe and assess in its complexity, let alone to do this precisely or express mathematically. Therefore, we may attempt to assess the cognitive complexity indirectly, by considering (the complexity of) what has to be understood. In particular, the (complexity of) understanding has three primary sub-aspects: attributive, behavioural and functional.

3.1.2.2.6.2.1 Attributive Aspect

Many simple models are just a set of values representing physical, mathematical, and other quantities or attributes of the modellee in the model. In order to be able to do this properly, the modeller has to know as much as possible about them – domain of the attribute, type of the possible values, frequency of change, *etc*. Insufficient knowledge or careless performing of this modelling task can lead to huge problems. An example is the Y2k-problem (also known as the millennium bug): in many software programs and even in some hardware devices the time was modelled in a way that caused overflow at the millennium change (December 31st, 1999, 23:59:59), leading to information loss and, respectively, to improper date display and interpretation. When interpreted by humans, the incorrect time-display is just annoying, but in the case of machine interpretation, which is needed for proper control, it can be dangerous.

Since most attributes are scalar, they are represented as numbers, and all that is discussed in Section 3.1.2.2.6.5.1 below should also be considered.

3.1.2.2.6.2.2 Functional Aspect

As already stated in the previous chapter, the (required) functionality of any model depends on its purpose. On the other hand, each function of an object increases its cognitive complexity, and each function that is to be implemented in a model increases its implementation complexity.

Instead of representing the functionality as the set of all (needed) functions, let us view it as the equivalent set of all (needed) elementary functions S_{EF} which can satisfy it (*i.e.*, a set of non-compound functions, having the same functionality). Now, the functional complexity C_F can apparently be represented as a function of this set $C_F=f(S_{EF})$ or as a function of the cardinality of the set $C_F=f(|S_{EF}|)$.

At some moment every elementary function of the modellee will be represented (or modelled) through a software function. The complexity of an elementary software function has, in turn, two sub-aspects. The first sub-aspect depends on the number of parameters (or independent variables), on the domain, on the co-domain[30] and on the mapping between the former and the latter. It can be called *exploitation-related* or *usage-related* or *external*, because it views any function as a black-box with certain number of inputs and outputs with an exactly defined relation among them – dependence of the outputs on the inputs and possibly also on the time. The second sub-aspect concerns only (the complexity of) the implementation itself, it depends on the exploitation-related sub-aspect and – possibly – also on additional requirements or restrictions. It can be called *implementational* (or *implementation-related*), since it is almost irrelevant after the implementation.

3.1.2.2.6.3 *Compositional/Combinatorial Aspect*

The compositional aspect applies to the complexity of both the modellee and the model. Apparently, it depends on the number and type of components as well as on the number and type of interrelations among them – *cf.* Figure 3.14. Whereas it seems intuitively clear that the complexity is directly proportional to the number of components, this is not always true, or at least not every aspect of complexity is directly proportional to the number of components. The "chest of nails" example from Edmonds (1999a, p. 44), shows that even a huge number of objects of the same type (*cf.* Figure 3.21) can seem simple "at least in some circumstances":

> It would be very unusual for someone, on opening up a large chest of nails, to exclaim "Oh, how complex!". This indicates that the mere fact that there are a great many nails is, at least in some circumstances, insufficient for the assembly be considered complex (this would not prevent someone modelling the forces inside the pile considering it complex).

On the other hand, even such a straightforward thing as counting of simple objects can be a difficult or "complex" task when they are not ordered – *e.g.*, try to count the screws in Figure 3.21.

A more careful analysis of this example shows that such "circumstances" can only be another name for purpose, role or other factors that influence complexity, as those in Figure 3.14, and play therefore a significant role. Given three different people for three different tasks: to get the nails from the chest, to hammer the nails, and to model the forces inside the pile. Each person would have totally different *impression of complexity* of the respective task. Since nothing else is specified in the above example, one assumes that all nails are of the *same type* – in the sense of material, form, size, outlook, *etc.*

---

[30] Earlier often referred to as range of the function.

Figure 3.21. A "chest of nails" (26 screws)

Now consider how the feel of complexity would differ if three chests of the same type, containing the same number but differently ordered nails, are compared: a) uniformly oriented nails (like the matches in a matchbox, one direction only; *cf.* Figure 3.22); b) nails oriented in two directions (*i.e.* some are rotated 180°; *cf.* Figure 3.23) and c) not oriented at all (*i.e.*, "chaotically" oriented, *cf.* Figure 3.24)? Of course, case c) leaves the impression of the highest complexity and case a) of the lowest, but they all are actually of the same complexity.

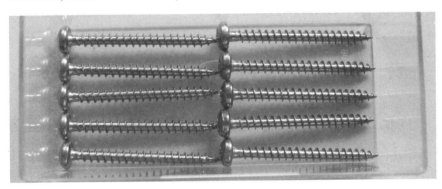

Figure 3.22. A box with 10 ordered screws

Figure 3.23. A box with 10 ordered screws (two possible orientations)

Next consider how the same persons' feel of complexity would change if we start to change the composition of the nails in the chest so that the amount remains the same while the number of nail types increases. The impression of complexity indeed gets much stronger! More precisely, the uncertainty of our information about the system of objects and its possible and actual states increases – this time in two dimensions.

Figure 3.24. A box with 10 unordered screws

To summarize, at least four factors influence the compositional aspect of complexity:
1. the number of distinguishable types of components;
2. the number of distinguishable components from each type and
3. the number of distinguishable relations[31] among them; the component's order (according to their orientation in space, or other criteria) is just one of the possible interrelations;
4. the number of distinguishable relationships from each relation.

In Goltz (2000) there is a comparison of the complexity of the models of bicycle, car, ship/aircraft, expressed by the number of their parameters – 10^3, 10^5 and 10^6, respectively – saying that "*only a subset of parameters is needed for sufficient collaborative product development – approx. 500 (0.5e03) for a car*". Unfortunately, the number of types of parameters is not given. Parameters can influence enormously the *flexibility* (*cf.* the previous chapter for definition). If the software models are viewed as non-modifiable building blocks or *components*, the flexibility of each component is apparently proportional to the number of its parameters: a component with no parameters is not flexible at all (*i.e.* rigid), while the more parameters are used for the component's definition, the higher its flexibility is.

3.1.2.2.6.4 Changeability Aspect (Dependence on Acquired Information)
Let us try to analyse how complexity depends on the available information about the matter. Assume that you have to read a text and you are trying to predict the

---

[31] The term relation is used here in the sense of the totality of all relationships of the same type – *cf.* Sowa (2001).

sense of the sentence currently read. At the beginning your uncertainty is infinitely high, because the (coming) sentence could express virtually everything. If you possess the respective background knowledge, there are no errors in the sentence and the sentence is clearly expressed, the chances are that after the end of the reading, and possibly a short time for thinking, you will know what the meaning of the sentence is. In other words, we could say that the complexity has dropped from infinite to zero during the reading. If we consider how the acquired information increases during the reading, we can see that the total amount of information is built up from at least four components: gain of information from every recognized symbol, word, phrase and sentence. With each additional piece of information the possible meaning of the sentence becomes more and more restricted, approaching the meaning intended from the author. The predictability increases and the uncertainty – respectively complexity – falls. Learning is a similar process, leading to decreasing of the imaginary complexity.

When reading a human's writing, there is little randomness in the stream of information. In other cases, though, the randomness can significantly increase the dynamics of the expected events and thus the resulting complexity of the respective system.

3.1.2.2.6.5 *Representational Aspect (Representational Complexity)*

As an idea matures and is elaborated to an archetype (*cf.* Figure 2.18 in Chapter Modelling Basics), the need to represent it increases more and more. Until a product or a process is derived from a given archetype, many different representations or models may be needed. Even in contemporary engineering, the first representation is very often a kind of a sketch, and the last is almost always a software model, run on a computer.

As already mentioned (*cf.* Figure 3.14), the complexity of a (software) model depends, among other things, on its purpose, on the complexity of the modellee and on the chosen representation. The representational complexity itself depends mostly on the model's purpose and on the implementation. Of course, sophisticated (future) products have numerous components which are usually of different types. Naturally, different types of components (or objects, or entities) are represented in a software model differently. So let us start with one of the easier complexity aspects and consider how some of these different types of components are represented in a software model.

3.1.2.2.6.5.1 *Numerical Complexity*

The computational complexity is probably the best investigated complexity aspect until now. One of the central fields of the *theory of computation* is even called "complexity theory" (*cf.* for instance Rosenhead (1998); Håstad (1999); or Goldreich (2000)) and is concerned with the study of the intrinsic complexity of computational tasks, tending to aim at generality. In this text the subset of the computational complexity, dealing with the representation of scalar values by means of numbers is called *numerical complexity*. Since in the end every software model is represented by means of numbers, the numerical complexity stays behind (or forms the basis of) all other kinds and aspects of complexity. But let us see why the representation of such a simple thing as a number can be complex.

Assume we have some sort of storage that is in a sense perfect and allows us to store *any* integer number in it. Since the number of the existing integer numbers is infinite, the probability that the storage has (or represents) certain number N is $1/\infty$, meaning that the uncertainty about the content (or state) of the storage tends to ∞. Therefore, the complexity of this variable after Definition 3.1 – let us call it numerical complexity and denote it with C_N – would be also ∞. Unfortunately, the possibility to represent any number from an infinite set of different numbers into a (digital) computer is technically unachievable. To surmount this problem the following measures are normally applied in the context of (digital!) computers:

5.   The numbers are represented according to a specially developed scheme that bounds the integer numbers to some limits. When real (or floating point) numbers have to be represented, in addition memory space is traded for precision (*i.e.*, each number being saved is rounded to the nearest exactly representable one).

6.   For each variable is reserved countable storage with respect to the expected limits for varying and precision of the representation; the amount is a multiple of byte (8 bits) and usually varies between 1 and 10 bytes (although higher values are also possible).

In this situation it can be assumed with sufficient accuracy that the representational complexity of any number depends only on the size of the reserved for it memory M, measured in bits, and no longer depends on the number's type:

$$C_N = f(2^M)$$
(3.11)

3.1.2.2.6.5.2 Text

Formally, any text representation in a computer[32] could be viewed as a sequence of as many codes as the number of symbols in the text. Each code is actually a number that can be directly stored in the computer's storage or manipulated as needed. Different *code tables* put each symbol of any (known to the platform) alphabet with the symbol's number according to the alphabet's order. Thus, the complexity of the placeholder for one symbol of text is equal to the complexity of the code (*i.e.*, of a number), used for the internal representation[33] of the symbol. The complexity of a text of length L (one-byte) symbols would be:

$$C_{text(L)} = L * C_N$$
(3.12)

Equation 3.12 just expresses the apparent fact that the complexity of any text is proportional to its length, but many people are shocked by the fact that the text complexity is not dependent on the text sense, semantic or "value". Frizelle and Suhov (2000), for instance, mention that Gell-Mann is not satisfied with Kolmogorov's definition of complexity:

---

[32] Only the representation of the text through alphabetical symbols and *not* the symbols' appearance in the sense of font, rendering, *etc.*, is meant here!

[33] Each code table establishes the relations between a certain number of codes and their visual representation. Some widespread code tables are ASCII (7 bit code, allowing representation of 128 characters), ISO (in different variations, but usually 8 bit, allowing representation of 256 characters) and Unicode (16 bit, allowing representation of 65536 characters).

He particularly objected to the fact that the complexity of a Shakespeare text was less than that of an unintelligible one generated by using random numbers.

Actually, the complexity of the text representation and the complexity of what is represented by the text itself are totally different things with totally different complexity! In the citation above they are mixed up, which has led to confusion! Of course, both the creation of Shakespeare's texts and their content have much higher complexity than that of a text "generated by using random numbers", but the complexity of their representation can be comparable or even equal in the sense of Equation 3.12. This confirms again how important is the proper understanding of the different complexity aspects and distinguishing them from one another.

3.1.2.2.6.5.3 Geometrical Objects

Geometrical objects are probably in third place in frequency of use after numbers and text. Since their representation relies on the representation of the scalars, apparently their representational complexity (not the visualization complexity!) will depend on the numerical complexity in the following way:

v) A scalar will be assumed to have the same complexity as the variable representing it.

w) A point or, respectively, a vector, will have different complexity depending on the number of dimensions it has: a one-dimensional point (point on a line or on the number line) has the same complexity as any scalar.

x) A two-dimensional point (on a plane) would have complexity:

$$C_{p^2} = C_N + C_N = 2C_N \tag{3.13}$$

y) Analogously, a three-dimensional point (space point) would have complexity as shown in the next formula:

$$C_{p^3} = C_N + C_N + C_N = 3C_N \tag{3.14}$$

z) If we express the same with the dimension D as a parameter, instead of 3.13 and 3.14 we can use:

$$C_{p^D} = D * C_N \tag{3.15}$$

aa) A line segment, represented as start and end point (on a plane or, respectively, in space), would have complexity (derived from 3.15):

$$C_{LS^D} = 2C_{p^D} = 2 * D * C_N \tag{3.16}$$

bb) By induction, the complexity of a poly-line (non-filled polygon) C_{PL} with k points would be:

$$C_{PL^D} = k * C_{p^D} = k * D * C_N \tag{3.17}$$

cc) The complexity of a two-dimensional circle, represented as a centre point and radius (*cf.* the model in Figure 2.34), can be derived from 3.15 and 3.11 and is:

$$C_{Circle^2} = C_{P^D} + C_N = 2 * C_N + C_N = 3C_N \qquad (3.18)$$

The complexity of a space circle cannot be derived from Equation 3.18 as simply as Equation 3.14 was derived from Equation 3.13. The problem is that it is possible to build infinitely many 3D-circles through a (three-dimensional) point and a radius – $i.e.$, the available information is insufficient to define the circle uniquely and unambiguously. Therefore, we need an additional element or restriction to avoid ambiguity – $e.g.$, two points on the circle, or a plane description, or a vector normal to the circle's plane. The latter option is most frequently used, since it is compact and in addition the normal can be used for some visualization purposes (in contrast to the additional points on the circle). Thus, the complexity of the 3D-circle's internal representation can be expressed as a sum of the complexities of a 3D-point, a scalar (the radius) and a 3d-vector (having the same representational complexity as a 3D-point):

$$C_{Circle^3} = C_{P^3} + C_N + C_{P^3} = 7 * C_N \qquad (3.19)$$

By induction it can be proved that the representational complexity of any software model of a given geometric object can be expressed as a function of the number of its attributes (or parameters) and the scalar complexity C_N.

3.1.2.2.6.5.4 Geometrical Aspect

Intuitively, the geometrical aspect of model complexity depends on the geometry of the model and on the geometrical relations among its components. In addition, it depends indirectly on the geometry of the modellee and on the chosen representation of the model. For software models of geometrical objects, though, at least two different representations are needed ($cf.$ also Section 3.1.2.2.1.1): the internal representation, used for (permanent) storing of the objects and for their (internal) processing, and an external representation or $visualization$ (see below), used to present the objects to the user and, possibly, to other systems.

Let us consider again the model of a circle, represented in Figure 2.34 of the previous chapter. The values of the leaf nodes of the structure represented there are sufficient for representing the model of any unique circle internally in some system. Actually, if during the saving operation these values are randomly "put" on the storage, there is no guarantee that by the next loading of the model (even in the same system) each value would go into the respective variable of the model, and it would not result in swapping, rotation or full mess of all values. Therefore, the consistency of the model (in the sense of proper relations among the values of the attributes/parameters and the respective variables) should be ensured by the software routines that write and read the model to and from the storage. This process is usually called $serialization$ and the routines implementing it – $class$ $serialization$ $methods$. In the object oriented programming the class-specific information, including all class methods, is saved separately from the instance-specific information, but only together these two different types of information allow for the use of the model, which is represented by means of numbers. Thus, the geometrical aspect of the representational complexity has influence on the visualization and on the handling of any (geometrical) object.

3.1.2.2.6.5.5 Visualization-related Aspect

The visualization plays a major role in every engineering discipline. No engineer can manage without some kind of graphical representation of his ideas. In practice, sketches, drawings, diagrams, flowcharts and many other kinds of visualizations are used. Due to this diversity it is difficult to introduce a universally valid measure for this aspect. It is obvious, though that the visualization complexity depends mainly on its purpose. According to the purpose, an adequate minimally necessary quality has to be specified, which is the factor of primary importance in determining what kind of visualization is required.

The second factor with a major influence on computer-based visualization is the art of the generated images – vector or raster images. Most contemporary computer displays and other devices for graphic output utilize raster graphics, but many existing devices use also vector graphics.

Every raster image (known also as bitmap) comprises a finite number of points (known as picture elements, or *pixels*), whereas every point has one of the number of possible colours (known as *colour depth*). Therefore, the visualization-related aspect of complexity of a bitmap can be represented as a function of the number of pixels used and the number of possible colours.

3.1.2.2.6.6 Handling Aspect

The handling aspect of complexity is relatively little studied. Therefore, we shall mention some factors that the handling depends on:

7. the purpose of handling
8. complexity of the subject to be handled
9. order of the components (if it exist)
10. the nature of the handling itself
11. the possible ways of handling
12. interdependencies and interferences of the above factors

3.1.2.2.6.7 Temporal Aspect

In many cases we know that a certain event will happen – *e.g.*, because it is recurrent – but it is difficult to predict the exact time of the next occurrence. For instance: we know that any tool will get broken or worn-out at some moment, but despite statistical information about the same tool type, the exploitation conditions are so different that it is impossible to predict the exact time. We know that the next bus will arrive on the bus-stop approximately as its schedule prescribes, but the exact time is difficult to predict.

The temporal aspect (or component) of complexity is difficult to measure, but we could say that it is proportional to the average frequency of the recurring event and inversely proportional to the deviation of the events from the average period between two occurrences.

3.1.2.2.6.8 Procedural Aspect

It seems that the *complexity of processing* of a group of entities (or also procedural complexity) C_{proc} – in the case of nails example, the complexity of getting, hammering or modelling – is directly proportional to the number of the types of entities to be processed t and to the complexity of each type TC:

$$C_{proc} = \sum_{i=1}^{t} TC_i \qquad\qquad (3.20)$$

Indeed, the processing of a given number of entities when they are stand-alone and when they are in a system is different. Each entity could either impede or make easier the processing of any other entity of the system. For a system of K entities this could be expressed by a two-dimensional matrix correction factors of size (K,K) where each element $cf_{i,j}$ is a factor, showing the influence of element i on the processing of entity j. Each $cf_{i,i}$ should be equal to TC_i, expressing no influence on the own processing, while all other factors can be 1 for no influence, have values between 0 and 1 for a facilitating effect and values greater than 1 for an impeding effect. Note that $cf_{i,j}$ is not necessarily equal to $cf_{j,i}$, nor to $1/cf_{j,i}$. The precise formula is still to be elaborated.

In turn, the *effort for processing* E_{proc} depends on both the number of entities of each type N_i and the effort for processing of each type TE:

$$E_{proc} = \sum_{i=1}^{t} N_i * TE_i \qquad\qquad (3.21)$$

Apparently, each TE depends on the respective TC in direct proportion (although the exact function will vary from type to type), so that we can write $TE = f(TC)$ and after substituting it in Equation 3.21 we get:

$$E_{proc} = \sum_{i=1}^{t} N_i * f_i(TC_i) \qquad\qquad (3.22)$$

Thus, we see that the number of types can have much stronger influence on the effort for processing than the number of entities/instances. The determining of the dependence of the TE on TC needs additional investigation. This leads to another aspect, tightly bound to the procedural aspect: the dynamical aspect of complexity.

3.1.2.2.6.9 Dynamical Aspect

Let us for a moment assume that it is possible to take a "snapshot", describing or reflecting everything about a given system in a given moment – number and type of components, their structure, organization, position and orientation in space, interrelations, *etc.* The whole information, related to this moment, is usually referred to as *state* of the system and can be viewed also as one of the systems models. When comparing snapshots, taken at different (moments of) time, shows no difference, the system is considered to be a *static* one. If the snapshots differ from one another, we have a *dynamic system*. Analogically we can define static and dynamic models; in this case, the analysed model is viewed as a model-system, and its snapshot can be compared with a meta-model.

Although it might happen that many snapshots have no discernable difference among them, it is quite probable that the system is not static, but rather in a kind of equilibrium for the time of observation. If the equilibrium (of the forces influencing the system) is breached by some event, the system becomes dynamic again. Another possibility is that the changes are simply not detectable or not measurable with the available methodology and tools. Thus, the term "static" is usually true only for a specific period of time.

Indeed, it is difficult to find "really static" systems in nature. A model of a system, however, can be static even when the modellee is a dynamic system. Such static models (the above mentioned snapshot is one example) can help us, for instance, to study time-independent characteristics of the system like structure, organization, relations, *etc*.

As already mentioned in the previous chapter, each process involves at least two objects forming a system, and implies some changes in the respective system. Therefore, it makes sense to use dynamic (*i.e.*, time-dependent) models when we model processes. According to the dependence of the chances on the time, two major types of processes can be distinguished *continuous processes* and *discrete processes*:

Definition 3.5: *the process, for which any two snapshots taken within time* Δt, *where* $\Delta t \rightarrow 0$, *are different (meaning that something changes continuously), is a continuous process.*

Examples of continuous processes (at least within the time period observable by humans) in nature are the flow of rivers, the movement of celestial bodies, *etc*. Examples of industrial continuous processes are many chemical processes such as galvanization, distillation, *etc*., some metallurgical processes and others.

Now suppose that we can take snapshots of a given process infinitely fast and during the time interval ΔT we have taken $N>1$ snapshots in equal time intervals $\Delta t = \Delta T/(N-1)$, denoted as S_1 to S_n, respectively. The following definition can be derived:

Definition 3.6: *Given the process Pr is observed during the interval* ΔT *and N snapshots are taken; if within the set of all pairs of snapshots with consecutive indices* $\{\langle S_i, S_{i+1} \rangle \mid i \in [1, N-1]\}$ *there are both pairs with equivalent elements* $(S_i \equiv S_{i+1})$ *and pairs with non-equivalent elements* $(S_i \neq S_{i+1})$ *of each pair, the process Pr is a* discrete *process.*

The time intervals during which the snapshots remain without change are called *states* of the process. Depending on the choice of Δt it is possible to have states with duration $d > \Delta t$. The changes from state to state are sometimes called *steps* of the process or *transitions* from one state to another. The following observations concerning the states and steps of a given process can be made:

13. It is possible that over a period of time, some states reappear and other do not.
14. Some processes can have only a small (or countable) number of states, other can have an infinite number of states.
15. The number of observed (or discernable) steps St of any process is always greater than or equal to the number of snapshots.
16. When $St > S+1$ some of the states appear more than once during the process flow.
17. Processes with an infinite number of states are difficult to predict and to model, except if some functional dependency (of the process output on time) exists.

18. For many processes one or two special steps – *start* (or *initiation*) and *end* (or *cessation*) – can be discerned.
19. Making the time of observation ΔT long enough, or extending it towards infinity, tends to discover that the process is neither continuous nor discrete, *i.e.* it is discrete-continuous.
20. If the output of an analogous process during the time of observation can be described as a periodic function of time, the process is considered to be a *periodic process.*
21. If a discrete process takes N different states during the time of observation ΔT and they recur always in the same order, the process is considered to be a *periodic discrete process.*

If a compound model is represented as a system of its sub-models (or components), its uncertainty will be proportional to the number of components involved and to the number of states each sub-model has. Therefore, the complexity C of a system built up from N components, each with the minimal amount of states (two – *e.g.*, active and passive), will be

$$C = f\left(2^N\right) \tag{3.23}$$

If the different components have a different number of states s, Equation 3.23 is modified to:

$$C = g\left(N, \prod_{i=1}^{N} f(s_i)\right) \tag{3.24}$$

A higher number of states per component leads to faster increase of the complexity. If a component is compound, the number of its states s_i is directly proportional to its complexity and can be calculated by applying Equation 3.24 – recursively, if necessary. If a component is defined parametrically, the calculation becomes more complicated. In general, the mere presence of a parameter already increases the complexity of the related component, but it is difficult to define or measure this increase. The particular value of the parameter does not always have influence: parameters controlling the metric do not change the complexity, but those controlling the topology or the structure of the component possibly do.

3.1.2.3 Measuring Complexity

We need to be able to measure the complexity at least for being able to compare it in different cases. On the other side, being unable to accurately measure something means insufficient knowledge about it. Insufficient knowledge, in turn, means incapability to control it, and lack of control means to be at the mercy of chance. Therefore, we look for a possibility to precisely assess and measure the complexity of any task, system, model, *etc.*, which we have to deal with.

Considering once again everything discussed until now about complexity, we can say that complexity is directly proportional to the number of components N, the complexity C_N of each component, the relations between components R_C and others.

$$C = f(N, C_N, R_C...) \tag{3.25}$$

Then the process complexity can be defined as the uncertainty of predicting the next event in a system or the next state (or the changes in the state) of an entity. According to its dependence on events the prediction can be of two types:

- event-independent – when just the time flows but nothing else changes, *i.e.* when the next change within a system is conditioned only by (the controlling module of) the system itself,   and
- event-dependent – concerning the response of a system to a given (already known!) event.

According to its dependence on external influences, the prediction can also be:

- system-internal, *i.e.*, concerning only the system itself, and
- all-embracing (or system-neutral), concerning either the immediate environment or the whole universe.

Now let us suppose that the destination is a human and the source is his environment as illustrated in Figure 3.25. Similarly to Shannon's approach it is possible to describe the probability with which a given event will happen in a given context (under context we shall understand some finite number of events that have already happened). For this reason we have to know all possible events, the probability of their coming and how this probability depends on the previous events (*i.e.,* on the context).

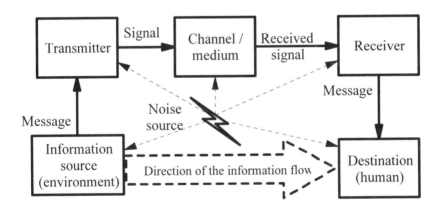

Figure 3.25. Model of the data transfer between a human's environment and the human himself, after the idea of Shannon (1948) for a transfer between a source and a destination

3.1.2.4 Improving Dealing with Complexity

Before looking for ways to reduce complexity, we have to know the answers to a couple of other questions, *e.g.*, "Which complexity do we want to reduce: design, use, maintenance?", "What is the price of the complexity reduction: reliability, flexibility, money or something else?". And we must not forget that the cause of too high complexity may be something useful – *e.g.*, too high functionality. Consequently, in most cases it is desirable not to reduce the complexity by reducing the usefulness, but to find better ways of dealing with it instead.

Let us first consider which complexity could be reduced and when this would make sense. Apparently, the absolute complexity of something real or already existing (object, product, process) cannot be reduced (but: if this is an artefact, we still could work on reducing the complexity of its next version!). The absolute complexity of a model, though, can be reduced by choosing either a less detailed representation or a more efficient/simple representation of this model.

Reducing the discernible complexity (even if possible) does not always make sense, since important details could be missed. It is possible, though, to reduce the complexity which has to be dealt with for solving a certain task. We shall call this part of the absolute complexity a *due complexity*. Since by definition one of the main purposes of each model is to allow us to abstract from unimportant details, we can assume that to achieve optimal (modelling) results, the *essential complexity of a model should be equal to the respective problem's due complexity*. The interrelations of different complexity types are represented as a Venn diagram in Figure 3.26. The absolute complexity is not represented, since it almost always can be considered infinite.

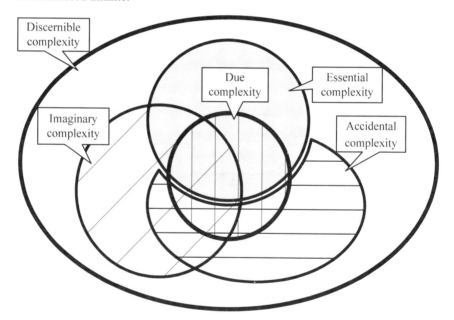

Figure 3.26. Complexity types and their interrelations

Luckily, although the due complexity is influenced by all other complexity types, its size remains the smallest – especially when the task is properly divided into subtasks and they are distributed among appropriate experts. Moreover, the due complexity can be reduced by means of several techniques (*cf.* Figure 3.27), but most important are probably two of them: the separation of different aspects to deal with (*e.g.,* authoring from use; *cf.* Table 3.5), and the reduction of the imaginary complexity by means of education and training. Even when the imaginary complexity is not reduced to zero, the level of the (person-dependent) critical complexity would be lowered.

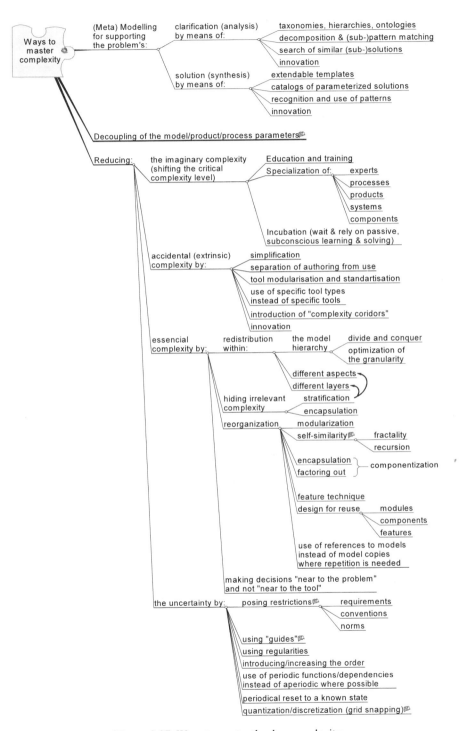

Figure 3.27. Ways to master the due complexity

Other important techniques concern encapsulation and componentization, complemented with conformity-conventions. These techniques reduce the due complexity and are well suited not only to software and software models, but also to parts of physical products or processes – we can view a gearbox, a suspension or an engine as components for a car.

It is important to develop appropriate methods and tools that would allow us to:

22. easily achieve encapsulation and componentization;
23. divide complexity into levels, so that each level can be served, controlled or managed by an average expert for that level;
24. change temporarily/locally the due complexity for supporting the decision taking;
25. selectively view different subsets of components of the model or links among them according to different criteria (aspect, level of detail, context, *etc.*);
26. visualize and manipulate software structures and other elements in a number of different ways (graphs, flow-charts, class diagrams, *etc.*)
27. split entities or models down to such level that the complexity of the respective sub-entities falls below the critical level;
28. reduce new problems to well-known and well-solved problems, leading thus to extended reuse of existing (part-) solutions and increased profitability.

As shown in Table 3.5, the design, production and use of an artefact usually have different and independent from one another complexities. This means that if these activities are separated from one another, the due complexity, associated with each of them will be distributed among different persons, respectively.

3.1.2.5 Complexity of Software Models

The complexity of software is a topic that is intensively elaborated by computer scientists. Nevertheless, most of this work deals with complexity from the viewpoint of a software developer or creator. Since the software does not age physically, a reasonably written and error-free program could not only be used forever, but also replicated and reused infinitely. If we consider the ratio between the *number of times a given software is used* and the *number of times it is created*, or alternatively, the ratio *duration of use* to *duration of creation*, it can be discovered that the higher this ratio is, the higher is the appreciation of the respective software. In such situations it becomes much more important to be able to measure the software complexity from the user's rather than from the developer's viewpoint. Since the use of any software is generally easier than its creation, the increase of this ratio can be viewed as a way to a better use of human resources.

Consider for a moment a software program or model as a black box with a specified number of inputs and specified number of outputs. Let us define the purpose of any software as *producing a (data) result by means of processing some input data*. Since the complexity of a software routine (or function, or procedure) from the viewpoint of its user does not depend on the internal states or structure of the model, and since by definition the output always depends in the same way on the input (*i.e.*, for each combination of input variables there should exist exactly one combination of output variables, independently from the number of input and output variables) we can say that:

- the software complexity from the user's viewpoint (in this case it is the same as the due complexity for the user) can be calculated without knowledge about the internal structure of a model or the way it works;
- the due complexity can be viewed as pure numerical complexity (*cf.* Section 3.1.2.2.6.5.1 and Equation 3.11) and depend on the number of inputs and outputs and on their (theoretically possible) information content.
- when there are no internal or global variables, influencing the way of working, and the inputs and outputs are independent from the time, the due complexity $_{due}C_{SWRoutine}$ is smaller than or equal to the greater of the sums of the numerical complexities of all input variables (each denoted $_{inp}C_{N_i}$), and all output variables (each denoted $_{outp}C_{N_j}$).

$$_{due}C_{SWRoutine} \leq Max\left(\sum_{i=1}^{K} {}_{inp}C_{Ni}, \sum_{j=1}^{M} {}_{outp}C_{N_j} \right) \tag{3.26}$$

Why do I use an inequality instead of an equation here? Let us illustrate this on the basis of an example routine – say the routine for calculating the trigonometric function sine. What is meant herewith is a routine for a digital computer. This routine has one input parameter and one output result. Usually one will use for both the input parameter and the result variables of the programming type *double precision* (often denoted simply as *double*) using 8 bytes or 64 bits of memory. Although the real input parameter can vary from minus to plus infinity, leading to infinitely many possible values, the respective variable can represent only a restricted number of them – exactly 2^{64} (*cf.* Section 2.4.2.3.1.2). Similarly, although the output of the mathematical function sine can take infinitely many values between -1 and 1, the routine in question would be able to represent just the same finite number of them 2^{64}. Now, recall that the sine function is a periodical function, meaning that there will be many different angles, differing by a multiple of 2π that have the same sine. Consequently, if we test the output values for each of the mentioned 2^{64} possible (and unique!) values, many of them will not be unique due to duplicates caused by the periodicity. Hence, the output values would never take all possible values representable by their variable, or in other words, the numerical complexity of the output of this routine is smaller than that of its input.

The user's due complexity is typically much lower than the due complexity of the developer for the same routine. Therefore, a reasonable question can be raised: does the ratio between these two due complexities have any meaning? Since each software routine is capable of either solving some kind of problem or performing some kind of task, we could define this ratio as *simplification factor*.

Now how does this apply to software models? As far as we are just (re)using ready models, the due complexity for their user could be calculated according to Equation 3.26. The so calculated due complexity can also be viewed as a functional or behavioural complexity, and should not be intermixed with the complexity of visualization, the complexity of a print or the complexity of other representations of the model!

The due complexity for the software model developer has to be estimated on the basis of evaluation of the computational complexity.

3.1.3   Integration-related Issues

3.1.3.1 Enterprise View on Integration
Isolated products, models or objects do not offer much value in the contemporary world. Sooner or later, no matter how big an entity is, a moment would come when it must be integrated with other entities to get added value. This has been very well understood everywhere in industry and academia. Let us look at two examples.

In Starzyk (2002) is represented a strategic model of the Boeing Process Council. Boeing is well known for its aircrafts. Boeing can be viewed as a typical original equipment manufacturer (OEM), with subsidiaries in many countries all over the world. The main idea of this model is that there are three categories *people working together* (*i.e.*, integrated) – employees, customers and suppliers. The people are carriers of the (different types of) knowledge, so they have to be supplied with process and technology, on the one hand, and with means for cooperative work, on the other hand, in order to achieve integration of their efforts.

The second example is a statement by the company BEA WEBLOGIC, well-known for its activities as system integrator. The citation below is an excerpt from the Rapid Business Integration 8.1 Datasheet, presented in BEA Systems (2004):

> *It's today's top challenge—integrating applications, data sources, business processes, and people to deliver on objectives for strategic business advantage.*
>
> *Unfortunately, the process of designing, building, testing, and managing integration projects takes way too long. Worse, once 'finished', the story doesn't end there: the solution is typically incomplete, too complex, or doesn't fully meet dynamic business needs. Why? Because business integration is more than just integrating applications one-to-one or drawing process flow diagrams. It is about providing a versatile environment for modelling, automating, and analysing business processes that access enterprise applications and enable business users to effectively collaborate.*

If everybody understands the role of integration, a logical question arises: Why is it not possible to manufacture integrated products instead of applying integration to separately manufactured parts or partial products?

3.1.3.2 Background
Modelling of complex products, systems of products and their production is typically done by multi-disciplinary teams, working in collaboration. Usually each partner or team is specialized in solving the problems, related to a given phase of the product's lifecycle, therefore CAx-systems of many different types are used throughout the whole lifecycle. Thus, the final modelling is achieved by integration of models created in these different CAx-systems. As a result, the integration involves a great deal of data processing, the largest part of which is conversion from one format/language into another. The following problems are clearly observable:

29. not every needed sub-model is convertible, which leads to quality losses;
30. not every convertible sub-model is really needed, which makes the efficiency questionable;

31. users trust the tools implicitly. In turn this leads to two other problems (32 and 33);
32. problems caused by the underlying tools are initially misinterpreted as own problems;
33. the users think (about integration) in tool-specific terms and context;
34. 29 and 30 lead to *data sharing* as an alternative to data exchange;
35. 31 and 34 raise (again) questions like:
- What is, actually, integration?
- How can integration be achieved?
- Is integration achievable in a tool-independent way?

The quality of a solution depends at most on the quality of the problem formulation. Therefore, returning "back to the roots" of the problem and its new analysis combined with a search for tool-independent and novel solutions can certainly be useful. Moreover, a comparison of the known integration methods in software development, mechanical engineering and electronics may lead to revealing synergy effects and to mutual benefits for each domain involved in the comparison.

3.1.3.3 Integration of Two Models: Possible Interpretations

3.1.3.3.1 Definition
Since integration (of elements into systems, of partial solutions into a complete solution, *etc.*) is a general problem, existing in almost every branch of science and industry, there are countless definitions for it. One of the recent and most closely corresponding to our understanding definitions is given in Lutters (2001) as "the facilitation of mutual cooperation and interaction between distinct functions in the manufacturing environment". In my view, though, the definition would be even more exact if it includes the relation between integration and its purpose, therefore, we use a slightly modified definition:

Definition 3.7:   *the integration of two or more (manufacturing) elements is the process of making them work on one and the same task or contribute to the achieving of one and the same outcome.*

How the integration will be achieved – whether they will be physically joined, or will obey to the same control, or just the results of their work will be joined – is an implementation question, which is of secondary importance for the end-user, but critical for the integrator and for the quality of the achieved integration. Therefore, it seems reasonable to perform a more thorough analysis and try to discover what are the inherent traits of integration and what they depend on.

3.1.3.3.2 Integration of Elements
For achieving good integration we need to know the traits of integration, to specify what aspects the integration can have, and to complete some possible classification of the integration types.

3.1.3.3.2.1 Integration Traits
Our investigation revealed at least twelve such traits, some of which can be viewed as input parameters of the integration process and used to control (or at least to

influence) it. They are depicted in Figure 3.28, numbered according to their approximate importance – the exact order of importance may vary and is case-specific.

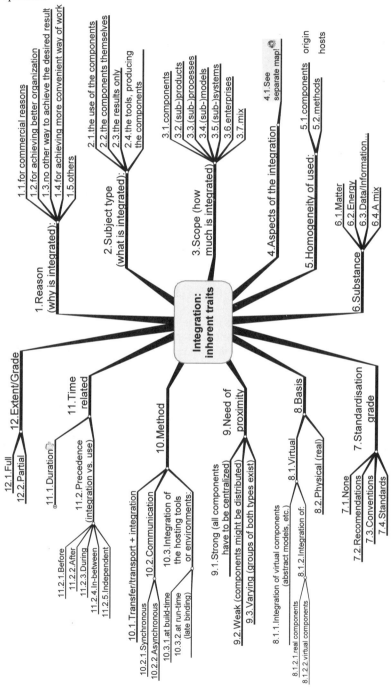

Figure 3.28. Most important traits of the integration

The first trait is related to the *reasons for integration*, which can be commercial, organizational, technical, *etc.* The second trait is the *subject of integration*, *i.e.* what exactly is integrated, since it is possible to integrate the components themselves, or the results of their work, or their use, or the tools producing them, or to have some combinations. The third trait concerns the *scope of integration*, or what kind of components are being integrated in terms of sub-products, sub-processes, sub-models, subsystems, enterprises or a mix thereof. The fourth is a composite trait and puts together various *aspects of integration*, most of which are case-dependent. The more important aspects are illustrated in Figure 3.29 and some of them are discussed in detail below. The fifth trait reflects the *homogeneity* of the involved methods and components, with regard to their origins and hosts (or environments). The sixth trait concerns the *substance* or character of the components that are integrated – are they material, energy, data, or something else. The seventh trait refers to the *standardization grade*: is the integration achieved by means of a standard, convention, recommendation, or none of them is involved in the solution. The eighth trait distinguishes real from virtual components and real from virtual integration. The ninth trait describes the *need for proximity* of the components to be integrated: have they to be close, may they be distributed and is it possible to have groups of both types. The tenth trait concerns the *method of integration*, *i.e.* whether certain components are transferred/transported, or it is based on (synchronous or asynchronous) communication. The eleventh trait is time-related, with two aspects: precedence – the relation between the time of the integration and the time of use of the components – and duration of the integration and use of the components. The twelfth trait concerns the extent of integration – whether it is full or partial.

Although the numbering of the traits sets them in a particular order of importance, this order should be viewed as preliminary and case-dependent. It is apparent, though that the trait groups with numbers 1, 2 and 3 have greatest influence on the (quality) of the pursued integration; the rest can be viewed as depending on the implementation.

3.1.3.3.2.2 *Integration Aspects*

Now let us return to the above-mentioned aspects of integration composing the fourth trait of integration. The most important/evident aspects are illustrated in Figure 3.29 (the numbering reflects the relative importance of the aspects, but is case-dependent). For any two components that have to be integrated, at least four of the thirteen mentioned aspects (*cf.* Figure 3.29) are applicable and have to be considered. It is apparent that the integration can have many "faces".

From the viewpoint of the end-user the functional aspect appears to be the most important, meaning that the product achieves its purpose. For instance, a person using a telephone to talk with somebody does not care how the device functions, neither how many other devices are involved nor how complex the system is that is formed together with the cable infrastructure. From the viewpoint of the engineer, though, the integration is more than just a means to an end, especially when the only way for the product to achieve its purpose is through integration. In this sense it makes a difference whether only the functions of the components are to be integrated or also the components themselves.

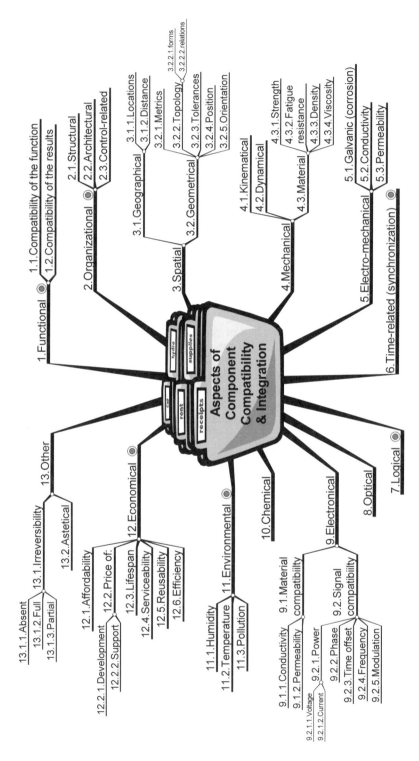

Figure 3.29. Aspects of integration. Aspects specific mainly to software components are marked with the symbol ◉.

The composition of several elements into a compound entity can be viewed as the classical or usual integration, especially when they are all of the same type – mechanical or electrical or software – or they all concern the same aspect. In such cases we can speak about *integration of homogeneous elements*, or *horizontal* or *extensive integration*. If there are at least two components of different types, though, we have to do with *vertical* or *intensive integration*. If the elements to be integrated or different logical parts of a component concern different aspects, we shall speak about *integration of different aspects*. Intensive integration is usually integration of different aspects. Example: mechanical (geometrical) and electrical aspects of a simple light switch (*cf.* Figure 3.30).

Figure 3.30. Electrical and mechanical models of one of the simplest mechatronical products - a light switch

Despite the extremely simple logical and electrical models, the mechanical model is much more complex. This is due to the need to implement auxiliary functionality in order to satisfy the safety requirements (isolation), esthetical requirements, environment-dependent requirements (protection against corrosion, humidity, dust, *etc.*) to solve secondary problems (suppression of sparks, fixing the wires, fixing on the wall, *etc.*). As a result, several aspects from those mentioned in Figure 3.29 are involved – functional, mechanical, electrical, electro-mechanical, *etc.* In addition, in this example the auxiliary functions are so interwoven with the primary function that it is almost impossible to design the individual aspects separately and to integrate them afterwards. In similar cases the best possibility for integration is to consider all involved aspects simultaneously during the design; in case of success the result is referred to as integration by design. However, the feasibility of such integration decreases exponentially with the increase in the complexity of the modelled/designed/manufactured product, as well as with the increase in the level of assembly in a hierarchically structured product. There are two reasons for this: a) after reaching a certain critical number of elements the complexity grows so much that it is impossible to consider all the involved aspects; b) the main method for reducing complexity is to divide and conquer, which is the opposite of integration. Therefore, more complex products cross many stages of integration, division and regrouping during their development. I argue that the optimization of this process is possible and that the most important areas for optimization are the *integration of functions/structures* and the *integration of aspects*.

3.1.3.3.3 Integration of Functions and Structures

Although not all integration aspects are applicable to software components (*cf.* Figure 3.29), their integration is also not easy. Nowadays the software is indispensable to the product and production development, as well as to numerous mechatronical products. Therefore, let us discuss some specificities of the software integration and try to find generally valid regularities that are (or can be) applicable to mechanical or other non-software components, too.

3.1.3.3.3.1 Analysis

Assume that the basic building block of the (information) systems serving a given manufacturing workflow, is called *micro information processor* (*μIP*) and contains input *i*, output *o*, environment input *e* and environment output *eo* as illustrated in Figure 3.31.

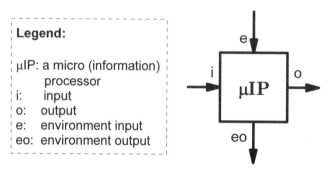

Figure 3.31. A black box model of a micro information processor

For now let us not specify whether a *μIP* consists of hardware, software or both: its most important property is that it can perform some processing (hence the name). Depending on which connections are in use (or are active), one can identify the respective blocks with specific mnemonic names, as given in Table 3.6:

Table 3.6. Typical *μIP* types (*cf.* Figure 3.31)

#	i	o	e	eo	μIP type
0	-	-	-	-	dead block, no real use
1	-	-	-	+	environment polluter (no real use?)
2	-	-	+	-	environment observer
3	-	+	-	-	generator
4	+	-	-	-	black hole (sink)
5	-	-	+	+	environment controller
6	-	+	-	+	noise/polluting generator
7	-	+	+	-	environment reporter
8	+	-	-	+	data-driven environment polluter
9	+	-	+	-	environment-driven black hole
10	+	+	-	-	transformer (see the text!)
11	+	+	+	-	repeater; programmable device (controller)
12	+	-	+	+	data-miner (see the text!)
13	-	+	+	+	controlled, learning generator
14	+	+	-	+	learning transformer (see the text!)
15	+	+	+	+	environment-aware μIP

Legend: A "+" denotes used connection, a "-" denotes an unused one.

The transformer (#10) deserves special attention, since it can be met most often, although in different variations. According to the definition of information processor in Avgoustinov (1997), three types of *IPs* can be distinguished depending on the relation between input and output sets of information: filter ($o \subseteq i$), generator ($o \supset i$) and converter ($o \approx i$). A special additional case of μIPs can be $o \equiv i$, meaning that the whole input is "repeated" without changes on the output. In this case either the μIP degrades to a pipeline and can be replaced with an arrow, showing the direction of the information flow, or we have a real repeater, which needs in addition a power supply from the environment (case #11).

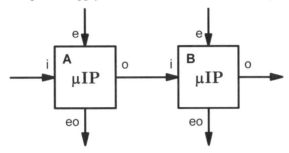

Figure 3.32. Serial connection of two μIPs

Now let us assume that the environment input and output e and eo are used for control and that the μIP is sufficiently intelligent to obey the instructions from the environment (via e) about what is to be done with the input data and how. A single instruction is usually referred to as a command, a sequence of instructions as a program. With this assumption it may turn out that μIPs like the transformer (#10 in Table 3.6) cannot function without a program and have to be replaced by μIPs like ##11, 12, 13 or 15. On the other hand, the environment observer (#2) seems to be of no use in this case unless it can accumulate the input and self-convert to some other μIP after reaching some threshold.

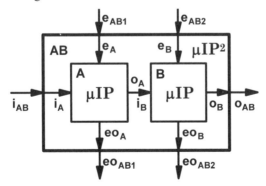

Figure 3.33. Virtual μIP, wrapping two connected μIPs

Let us consider how such blocks can be integrated (or put to work together). If the output of given μIP is compatible with the input of another one, it is possible to join both connections and thus also their respective μIPs. Connected in this manner, the two *μIPs* from Figure 3.32 can be represented by a virtual *μIP*, as illustrated in Figure 3.33.

In turn, the virtual *μIP* in Figure 3.33 could be represented in a simpler way by an equivalent *μIP* from the *second level* of the (modelling) hierarchy as illustrated in Figure 3.34:

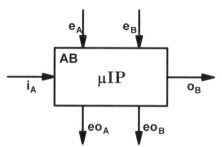

Figure 3.34. Equivalent of a serial connection of two *μIPs*

Similarly, two *μIPs*, connected as in Figure 3.35 (left), can be represented by their equivalent, shown in Figure 3.35 (right).

Since the name (input, output, *etc.*) of a *μIP*-connector determines its function, but not the type of the exchanged signals, it is at least theoretically possible to connect (by assuring compatibility of the signals) *o* to *e* and *eo* to *i*. In this manner it is possible to use the output of one *μIP* to control (program) another *μIP*, or to process program output (part of *eo*) of one *μIP* on another *μIP* as data (or input). At this moment we do not see any reason for connecting an input of one *μIP* with an input of another *μIP* – *i* with *i*, *e* with *e* or *e* with *i*. The same applies to the connecting of outputs – *eo* with *eo*, *eo* with *o* and *o* with *o*.

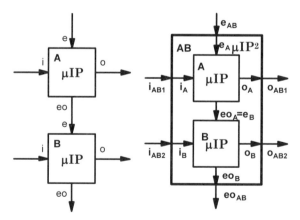

Figure 3.35. "Parallel" connection of two *μIPs* (on the left-hand side) and the virtual wrapping *μIP* (right-hand side)

On the other hand, it is possible to integrate two *μIPs* with no connection between any of their connectors – *cf.* Figure 3.36. This kind of integration can be viewed as purely mechanical integration and is used mainly to facilitate the

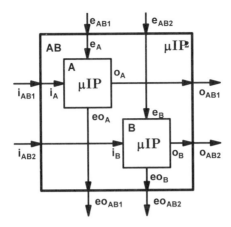

Figure 3.36. "Double-parallel" connection of two μ*IPs* and their equivalent

handling of several components at once. It has probably the widest use in electronics.

In general, the second-level μ*IPs* can be integrated again with other μ*IPs* from the first, the second or other levels and then form processors of higher levels – *cf.* Figure 3.37.

We shall refer to such compound components as *information processors* or *IPs*.

3.1.3.3.3.2 Classification of (Micro) Information Processors

Now assume that we integrate several μ*IPs* from the simplest usable type – transformer (#10 in Table 3.6) – into a higher level IP. An example with several μ*IPs*, connected in different combinations, is presented in Figure 3.38.

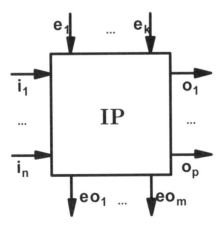

Figure 3.37. Higher-level (multivalent) IP

According to their connection types the components of any IP can be classified in three groups: internal (like *A3* in Figure 3.38), external (like *A7* in Figure 3.38) and "mixed" (both internal and external, like *A1, A2, A4, A5, A6*).

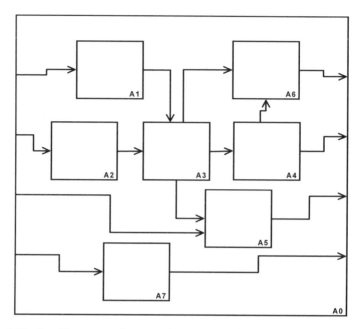

Figure 3.38. Specific types of connections among the components of a compound information processor

It should be noted that internal components reduce the externally discernible complexity of every component used as a black box. As external components may theoretically exist also as autonomous components, they are the first candidates for factoring out. Components that have both internal and external connections are often referred to as interface components or adapters (see below).

According to the number of its connections a given component can be singly, doubly, trebly or multiply connected. The singly connected components are untypical within a compound block, since they are either useless (see the discussion after Table 3.6) or could exist as standalone modules. Doubly connected components are the most widespread type. Trebly connected components are the nearest approximation of the μIPs. The multiply connected components are used in all other cases.

According to their (primary) function the modules or components of a compound model can be classified in at least five categories: connector, transformer, processor, controller and adapter.

3.1.3.3.3.3 Connectivity-dependent Traits of μIPs

Many traits of μIPs either depend on or are related to the number and type of connections. Here are some examples:

- flexibility as a function of the number of connections;
- productivity as the number of useful outputs;
- efficiency as a relation between the outputs and inputs;
- environment-friendliness depending on the type and number of outputs;
- interchangeability as an inverse function of the number of connections.

3.1.3.3.4 *Integration of Aspects*

For simplification reasons let us discuss this topic on the basis of models and see whether the conclusions are applicable to real objects. Our investigation shows that it is easier and advantageous to model the individual aspects separately and then to integrate the resulting models. Assume that the modelled object has N aspects, represented by respectively named rectangles in Figure 3.39, and the model of each aspect has a different number of properties, represented by different geometric shapes. Note that some shapes exist in all aspects – *e.g.*, the round shape denoted by *5*, *a* and *e*; we shall call such properties *common properties*. Some exist in only one aspect, *e.g.*, shape *b* in *Aspect$_i$*; these properties can be called *aspect-specific*. Some properties exist in several, but not in all aspects. And finally, there exist properties *specific to the result* of the integration of all aspects. They can be related to common properties – like *2*, which is related to *5*, *a* and *e* in Figure 3.39 – or to other types of properties – like *1*, which is related to *4* and *9*, but to no properties of other aspects in Figure 3.39 – or may arise from the integration and be specific only to its result – like *3* in Figure 3.39. These kinds of properties, together with the common properties, form the *core properties* of any compound model[34].

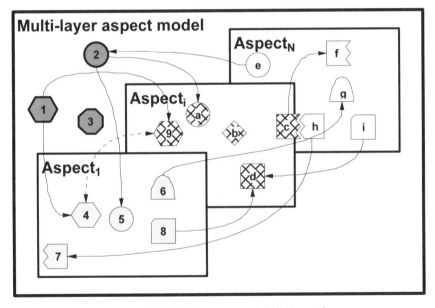

Figure 3.39. Models, aspects and their integration

Apparently, the integration of any individual aspect to the model depends on the percentage of common properties, but should/ought to rely on the core properties and has to consider all properties that are related to functions of primary importance. The full integration itself (*i.e.*, the integration of all aspects into the

---

[34] Similarly to the Kern-Features (translated to English *core features*), mentioned in Spur and Krause (1997, pp. 177, 510, 511), the core features also play a crucial role throughout the whole model lifecycle.

model) has *to ensure access to all properties[35] from all aspects* either explicitly or implicitly, otherwise it would not fulfil its purpose.

3.1.3.3.5 Componentization

According to our investigation one of the best methods for achieving integration of functions, aspects or both is the *stepwise, hierarchical, component-based integration*. According to the model centred approach the simplest models (*i.e.* those from the lowest level of the hierarchy) have to be implemented as components in the sense of the definition of the object management group (OMG): "*A physical and replaceable part of a system that conforms to and provides the realization of a set of interfaces*", *cf.* Booch *et al.* (1999). However, on the one hand they do not have to be "physical" (*i.e.* they can also be software components or other immaterial components) and on the other hand they have to be in addition:

21. self-contained, clearly identifiable artefacts that describe and/or perform specific functions and have clear interfaces, appropriate documentation and a defined reuse status Sametinger (1997);

22. software units that are context-independent both in the conceptual and the technical domain Ciupke and Schmidt (1996);

23. building blocks, which can develop independently from the container application; in contrast to Stal (1997), they can but do not have to be binary.

Finally, components can be encapsulated into other components, forming thus different levels in the model hierarchy.

Experience shows that best results are achieved when the components are formed according to functional criteria (and not, *e.g.*, implementation-related or other considerations). For software models it is reasonable to keep the functionality and the structure of each component as close to those of the modelled object as possible. The question is whether there is a level of componentization (or integration) hierarchy which is optimal for performing the intensive integration. Although we still do not have a final answer to this question, we can be assured that the lowest level (the level of the elements) is seldom well suited to intensive integration. Despite strong case-dependency of the integration (strategy), it seems that in general – starting bottom-up – several steps of extensive integration have to be followed by one or more steps of intensive integration and then the cycle repeats upwards until the desired result is achieved.

3.1.3.4 Integration-related Issues: Conclusion

Although the presented analysis is very simplified, it allows us to make some important observations. Probably the most important one is that the best way to achieve integration is to consider it during the design. Integration by design resembles other so-called *design for x* (DFx) and already well-established approaches by the fact that it also shifts part of the development effort towards its early phases. It shows how the following of simple design rules can allow one:

• to reduce the complexity of the components and their models;

---

[35] For completeness, assume that a function can also be viewed as a property.

- to reduce the effort for integration without compromising other design goals;
- to increase the component compatibility, integrability, efficiency and understanding.

It is important to note that the proper choice of scope, level and time of integration (with respect to the development cycle) plays a significant role in reducing the development effort. The defined traits and aspects of integration seem to be generally valid and usable for both nonmaterial (*e.g.*, software) and material components.

Another discovery is that probably the most important factor for the technical and economical efficiency of the components, as well as for their integrability, flexibility, reusability and overall qualities of the systems built up from them, is their *granularity* (average component size).

Electronics is a nice example of an area where systems with components extremely different in size and level of integration are used: of one end of the scale are the discrete electronic components such as resistors, capacitors, *etc.*; in the middle are integrated circuits (ICs) with high levels of integration, then come the ICs with very large-scale integration (VLSI), and at the end also, ICs with giant scale integration (GIS). The higher the level of integration of the components, the higher the grade of automation they offer, but at the same time their usability for purposes differing from the originally foreseen one, rapidly decreases. Thus, the need for flexibility is one of the main reasons why discrete electronic components are still produced.

The large majority of CAx-systems and other contemporary software tools (even together with the produced models, which are typically much smaller) lead to granularity that does not allow optimal solutions. The CAx-systems resemble GIS ICs and their flexibility is similar. Therefore, we should reconsider whether all tools belong to the solutions they create, *i.e.* whether they shall "live" together with these solutions or not. For that reason, the simple idea of shifting the focus of the product and production development from the software systems/tools onto the models themselves can lead to increases in the overall efficiency. This shift can be achieved by the placement of product/model specific functions (including those responsible for communication and integration with other components) into the models themselves instead of into the surrounding software system. In this way the models can be made flexible, long-living, intelligent and autonomous, which would improve their flexibility and the overall efficiency, too. It is much easier to integrate well-designed components than to integrate the systems used to create/host them. Along with the redistribution and reorganization of the efforts to achieve the same purpose, the main advantage of such an approach is the change in the way of thinking: to focus on the task and on the result to be achieved rather than on the tool to be used. The major achievement of such an approach should be a world of models, where all models can be integrated by design.

3.2 Problems Specific to the System Centred Approach

3.2.1 General Observations

> *Technology presumes there's just one right way*
> *to do things and there never is.*
>
> Robert M. Pirsig

A careful analysis of the SCA shows that in general it has three characteristic phases having the following goals:

dd)to determine the domain of the task;

ee) to find a CAx-system, well suited to model tasks from the (target) domain;

ff) to use the system for modelling/solving the task.

Thus, a CAx-system stays at the centre of any initiative, action and development. Removal or replacement of the system can have crucial consequences for the development process or in the worst case even break it. In addition, the following observations can be made:

- CAx-systems are comprised of numerous relatively independent *subsystems* or *modules*.

- Each module contains one or several *programs*, closely related with each other and dedicated to solving tasks in a sub-domain of the system domain.

- The modules communicate with each other either by means of direct procedure/function calls (APIs[36]), by common memory or – rarely – by files using (internally) standardized format.

- The programs within a module communicate typically by means of APIs.

The CAx-systems can communicate with each other by means of: a) data exchange; b) dedicated applications exploiting their API; or c) dedicated communication modules. Data exchange by means of a database or a PDM system can be a special case of b), c) or even a). Modules realizing a combination of the cases c) and a), *i.e.* modules dedicated to preparing large amounts of information to be suitable for other system types and to communicate the prepared information to the target system as files are often referred to as interface (or import, or export) modules.

CAx-systems typically expose *only a subset* of the internally known APIs to external applications. This is achieved by: a) simply restricting the API-documentation for the end user; b) exploiting implementation dependent mechanisms for access restrictions or c) a combination of the both. The aims of these restrictions are: d) keeping the integrity of the system and achieving thus its better stability; e) protection of the producer's know-how (against reverse engineering, *etc.*); f) additional profit (APIs are often sold together with additional software tools as *software development kits*).

---

[36] API is abbreviated from Application Programming Interface. It is a detailed description of the names of important functions and their parameters, allowing one to write new programs that use these functions.

In the case of data exchange (0.a) CAx-systems typically offer at least two different formats. Alternatives are the own format, some established standard format and the format of some other established CAx-system.

When the CAx-systems within a given software package have unrestricted (or at least extended) communication capabilities with each other, they offer to the end user comfort similar to that of an integrated CAx-system.

The peculiarities of the use of CAx-systems either lead to changes in the way the enterprises plan and work or require some additional tasks. In different situations or stages of the product development and production this could sound similar to the following:

- Instead of concentrating on solving design problems, the designer has to decide which of the available CAD systems is best suited to the current requirements.
- Instead of using the (CAx-)system he is accustomed to, the expert has to ensure compatibility with the rest of the production chain first.
- Instead of extending the CAD model, *e.g.*, with planning and/or manufacturing information, the next user of this model in the production chain first has to ensure readability and compatibility of the input information.

The experts of modern enterprises are expected not only to "*know how*" a given problem can be solved but also

a) to "*know which*" system Sys_x is best suited to solving problems of the given type;

b) to have a working copy of Sys_x and

c) to have expert(s) in Sys_x.

The requirements b) and c) and even a) could be "avoided" if the enterprise "*knows who*" can provide the respective (paid-) service for solving the problem – in such cases the respective tasks can be outsourced to this external service provider. The ready solution, coming back from the service provider in the form of a CAx-model, though, typically requires the same CAx-system in order to be used. Just a couple of years ago the so-called CAx-*viewers* (applications, offering preview of CAx-models without possibilities for editing) were offered. But they could not satisfy the user requirements for the prepared model. Therefore, condition b) holds in most cases, even for outsourcing.

If the diversity of problem types and CAx-system types is considered, it becomes obvious that production management has to deal not only with *technical* (achievability, quality) and *ergonomic* (comfort) aspects but also with the *financial* one (how expensive the solution is).

Let us reconsider some known facts, as well as unsatisfied industrial requirements and try to get to the root of the problems.

Each CA-system is designed to support a certain phase of the product lifecycle, but different requirements can apply to CA-systems serving the same phase in different branches. Thus, we can speak about orientation to a certain phase of the product lifecycle.

Models can "live" (*i.e.* be used) only within a system of the same type as the originating system – *i.e.*, they are *not autonomous*.

If a compound model is needed, every sub-model created in a "foreign" system must be converted in order to be integrated.

The number of existing system types is huge and continues to increase, along with the average system and model sizes.

Globalization poses new requirements for data access in a heterogeneous, multicultural environment. Due to its nature, however, the SCA is incapable of providing for optimal integration and lifespan of the models.

3.2.2   Problematic Issues

One of the worst disadvantages of the SCA is that the models inherit the problems of the tools, used for modelling – the CAx-systems. Among the most important problems are the lack of autonomy of the models created by means of a CAx-system, as well as the accidental complexity resulting from it.

3.2.2.1   Usability of a Model

A model that is not usable has no value. A reusable model has greater value than a "once-usable" model. A model that is usable for other purposes in addition to its original purpose has added value. Therefore, the usability of a model is a key characteristic, which depends on several prerequisites. Among them are the model's lifespan, its reliability, its dependence on other elements of the environment, such as host and energy, and the availability of qualified users that are capable of using the model. Some of these factors in turn depend on other factors. For instance, in the worst case the model's lifespan can be as short as the shortest lifespan of its components if the respective component is irreparable and not exchangeable. If the lifetime of the software model's host is over, and there are no more hosts of the same type, then the model cannot be brought back to life either. If the host needs energy or some other kind of supply (gas, coal, *etc.*), which is no longer available at that time or place, the model is again unusable.

An important precondition for a compound model to be usable is its integrity, *i.e.*, the simultaneous accessibility of all its components at the needed moment. This is achieved by means of different types of permanent or temporary integration of the comprising models.

3.2.2.2   Reasonability for Use of a Given Tool

Let us define the ratio of essential complexity (*i.e.*, the complexity of the respective model) to accidental complexity (in this case, the complexity of the needed tool - CAx-system, editor, *etc.*) as *reasonability to use certain tool* or as its *acceptable complexity* as follows:

$$acceptable\_complexity_{tool} = \frac{complexity_{modell}}{complexity_{tool}} \tag{3.27}$$

In the case of extremely simple models or modification types, where their complexity tends to zero, the acceptable complexity would also be zero, *i.e.* such a tool is unacceptable. In contrast, the lower the complexity of a given tool is, compared to the model's (expected) complexity, the more reasonabile it would be to use such a tool. Consequently, CAx-systems are not the best tools for handling small models.

3.2.2.3 Integration and Communication Problems

As shown in branch 2 of Figure 3.28, there are at least four ways to achieve integration: it is possible to integrate the components themselves, or the results of their work, or their use, or the tools producing them, or to have a combination thereof. Of course, each of these methods has its advantages and disadvantages.

The SCA utilizes mainly two methods for model integration:

24. integration of the models as illustrated in Figure 3.6;

25. integration (fusion) of the respective CAx-systems.

The disadvantages of method 24 are that it requires transfer and conversion (when the models originate from different CAx-systems) of at least one of the models. Any transfer leads to time losses; conversion leads to additional time losses, as well as to quality losses. Moreover, conversion is not always possible.

Method 25 means that the CAx-systems start to use the same internal format after their integration, which requires no model conversion and leads to better quality of the resulting compound models. Therefore, it seems to be better, in the long run, when models of a given pair of CAx-systems have to be integrated again and again. But this is not always applicable, and when it is applicable – much more effort is required. Even if we assume for a moment that the integration of two CAx-systems is always feasible, considering the great diversity of available CAx-system types reveals that we should not expect the emergence of The Integrated CAx-System uniting all CAx-system types (like the earlier idea of computer integrated manufacturing or CIM), which would solve all integration problems.

Now consider the fact that nowadays the size of a CAx-system is typically 10 to 1000 times greater than the size of any model created with it. Since the size is an indirect measure of complexity, and since each of these models lives in the respective CAx-system, the resulting ratio of essential complexity (the complexity of the model) to accidental complexity (the complexity of the CAx-system) is very low, meaning very low efficiency. For this reason it should be clear that there is not much room left for the integration of systems – at least not in the mentioned sense of system fusion.

As the use of standards for facilitating or avoiding the conversion of models does not show or promise much success (*cf.* 3.1.1.12 above), the integration of models with different origins by the SCA often remains a real problem.

3.3   Hypothesis

> *A fact is a simple statement that everyone believes. It is innocent, unless found guilty. A hypothesis is a novel suggestion that no one wants to believe. It is guilty, until found effective.*
>
> Edward Teller

> *The more you study, the more you know. The more you know, the more you forget. The more you forget, the less you know. So why study then?*
>
> Unknown author

In mechanical and electrical engineering, as well as in mechatronics, the SCA is already a well-established approach to the creation and use of sophisticated models. Nevertheless, due to the required multi-level and multi-aspect modelling, *the SCA causes imbalances and disproportions, which lead in the long run to high accidental complexity, non-optimal granularity, inefficiency, inflexibility and eventually – to extra model costs.* The most critical property of every CAx-system seems to be its complexity: the author has unofficially inquired for years among many designers about an expert knowing more than 10% of the CATIA's capabilities: without any success[37]. If we consider how much is invested in CAx-system training, it turns out that the system consisting of a man and a CAx-system has a not-too-high modelling efficiency! On the other hand, there is a demand that knowledge of the problem domain and knowledge of problem solving techniques should be as deep and complete as possible. This means that the accidental complexity of creating a model by means of a CAx-system is much higher than it should be.

Among other reasons for the mentioned imbalances are the irregular granularity of SCA, as well as the uneven use of the different data types.

The most important requirement for a product is that it has to serve/fulfil its purpose. The more functions a given product has, the more complex it is. Each required function increases the costs of the product's development and production, therefore, the buyers of the product compare competitive products and look for an optimal compromise between needed/desired functions, available functions and their price. There is a contradiction here: on the one hand, the excess functionality of any product increases its flexibility, and since at the time when a tool is procured not all possible applications are known, flexibility is very welcome, especially in SMEs. On the other hand, nobody is willing to pay for a functionality that is not going to be used, even if it is excellent! So how to reconcile these

---

[37] The inquired people were not only able to use the system, but were in a position to accomplish sophisticated design tasks with it.

contradictory requirements? In Warnecke (1996) the situation sketched here is described as "Corrections on the existing are not sufficient anymore, and troubleshooting is not the solution."[38] This view seems to be shared by many researchers, because numerous attempts to develop novel or alternative modelling methods or at least to improve important traits of the conventional SCA have been made. Some of these are discussed in Section 4.3.

At first glance it seems that the key to a solution, successful in both functional/technical and economic aspects, is the optimal choice of functionality. Yet, despite modularization and countless (module-) configuration and customization possibilities when buying a CAx-system, the choice of functionality is only possible within certain limits (*i.e.*, the granularity is relatively high). The smaller the needed functionality (think of the needs of a SME for a short period of time), the higher is the probability that even the system with the lowest functionality on the market will be over-dimensioned and possibly unnecessarily expensive for the purpose. As a result, the economic viability of a project can decrease rapidly. A much better solution from this point of view can be achieved by offering *functionality on demand* – an example is the implementation of a useful function as a (web-)service. Combined with ideas like *utility computing* (a.k.a. *computing on demand*) on the underlying infrastructure level, such an approach can lead to mind-boggling results. We shall discuss in the next chapter whether and how could such an approach be realized.

Having considered in addition the problems from the previous sections, it was decided to seek alternative approaches to modelling, which would satisfy the above-mentioned idea and would also be suitable for SMEs. The discovered discrepancies and problems, on the one hand, and the available experience, on the other hand, have led to a new concept for development, organization, use and lifecycle management of product models, which is especially interesting for SMEs. But first we need to formulate the requirements a hypothetical perfect (or ideal) modelling approach.

---

[38] In original: "Korrekturen am Vorhandenen reichen nicht mehr aus, und Fehlersuche ist nicht die Lösung."

4

Towards Better Product and Process Modelling

If we wish to make a new world we have the material ready. The first one, too, was made out of chaos.

Robert Quillen

4.1   Understanding the Aims

We must be systematic, but we should keep our systems open.

Alfred North Whitehead
Modes of Thought

Software producers and software users understand quite differently the purpose of the software systems which support modelling. The former typically try to develop systems offering rich functionality, while the latter are concerned about properties of the models designed with these software systems, such as lifetime, integrability with other models, compatibility, *etc.*

Now recall that most CAx-systems for mechanical, electrical or mechatronical engineering produce only models that can live in the respective creating systems, so that the models form, together with the creating system, a larger compound system. As a rule, these CAx-systems are not open, *i.e.* their users cannot (easily) modify them if an adaptation is required. This leads to two problems not mentioned until now: (i) the flexibility of a compound system is very low (*cf.* the definition of flexibility in Chapter 2), and (ii) the lack of openness is comparable to the circumstances of a mechanical or mechatronical or simply physical system for which there are "no spare parts available". So, the lifetime of such a "closed" compound software system is only as long as that of its component with the shortest lifetime. Even if the theoretical lifetime of the software components is infinite, due to lack of physical wear, their real lifetime can be extremely short, due to outdating or newly arisen customer needs.

The worst possible scenario is to have to use (systems of) models whose lifetime depends on the lifetime of the authoring tools. Such compound systems of model and tool cannot be viewed as agile. Yet, this is exactly the case with the majority of CAx-systems in the area of mechatronics, which is far from what can be regarded as perfect.

4.2   Requirements for the Perfect Modelling Approach[39]

> *Everything should be made as simple as possible, but not simpler.*
>
> Attributed to Albert Einstein

Sometimes there is a routine in our thinking, which may hinder us in noticing evident solutions of banal problems. With sophisticated problems it could be even worse – in some cases we seem reluctant to even think about a better solution, if a working one exists and is well established. Experience shows that in such cases it can be very helpful to forget for a moment all existing solutions, assume that from a technical point of view everything is theoretically possible, and try to formulate requirements for a solution that would best serve our needs – the perfect solution.

There is a saying that detecting the cause of a problem is already half of the solution. In this sense, defining the requirements (for a solution, for a product, *etc.*) can be compared to a proper splitting down of a problem or task into easily solvable sub-problems or sub-tasks.

Some experts view the elicitation and analysis of the requirements, together with the assessment of available means and the analysis of posed restrictions, as the most important tasks on the way to finding an adequate solution of any problem. In Kotonya and Sommerville (1998) the requirements engineering is described as a spiral process, whose sequential components are *requirements elicitation, analysis, negotiation, documentation* and *validation*. Further, they divide requirements elicitation into customer requirements elicitation and product requirements elicitation, and requirements analysis into requirements quality analysis and suitability analysis.

Many authors and theories insist, though that even before the requirements are specified we should consider the user needs or demands. The requirements can be derived then from these needs and demands. In the axiomatic design theory as presented in Suh (2001), for instance, four domains are distinguished – *customer domain, functional domain, physical domain* and *process domain*. Each of them has its characteristic vector, representing customer needs (*CNs*), functional requirements (*FRs*), design parameters (*DPs*) and process variables (*PVs*), respectively. The axiomatic design – which can be viewed as a kind of generic

---

[39] Even if it is arguable whether there is only one perfect approach or there can be more, notions like "perfect" and "ideal" have no comparative degree.

problem solving process – goes then iteratively from left (the customer demands) to right (the process variables), trying to formulate requirements that would satisfy the needs, then to find parameters (or parameter values) that would satisfy the requirements, and finally to find process variables (or their values) that affect the corresponding design parameters in a way to achieve the desired results. But if we discuss the modelling methodology and the respective tools in general, and not in a specific case, it is a bit problematic to determine particular customer needs. This means that even if we carry out a survey about the needs and desires of the (potential) customers, the answers would be so different that they would have to be reorganized and probably reformulated and grouped in an observable list of generalized needs. Therefore, a quicker and cheaper possibility is to conduct a mental experiment and to consider what would these needs and desires be. Suppose we start with just one customer need, which is

CN: get/have perfect models.

To satisfy this need it is sufficient to define just one requirement:

FR: use/apply a perfect modelling approach.

It seems simple but does not change much in the initial situation, because no approach until now is perfect and it is still unknown what would be a perfect approach. Therefore, we have to do some decomposition and try to find other relations. Although according to the axiomatic design CNs do not always need decomposition, in this case it would help us clarify the requirements. So, let us consider what it means for a model to be perfect. Of course, everybody would have his own preferences, but we shall say that perfect models have the following five main inherences.

The most important one is that each model should *resemble all important modellee traits* as closely as possible - CN_0. This is necessary in order to ensure that any *use of a model instead of its modellee* would be sensible at least for the purpose for which the model is created: CN_1. The purpose could be the creation or the improvement of the modellee itself, the planning of the modellee's production or simply the taking of a (perhaps more general) decision, related to the modellee. The next inherence of a perfect model is its *unrestrictedness*, *i.e.*, the possibility to model or represent anything without exception: CN_2. Next important trait to be expected can be *ease and comfort* when dealing with the models – model handling should be effortless: CN_3. Of course, the models should also be *robust and reliable*: CN_4. And last but not least – the perfect models should *live forever*: CN_5.

Now let us see what requirements should be imposed on the perfect modelling in order to cover these needs? We shall give the same indices to the corresponding requirements in order to keep the correlations clear and easily maintainable. Therefore, CN_0 leads to the first requirement – *allow representation of arbitrary attributes*: FR_1. CN_1, respectively, leads to the next requirement – *ensure the model adequacy*: FR_2. The requirement correlated to CN_2 still cannot be formulated clearly at the moment, but we can try again after formulating its sub-requirements, which correspond to parts of the decomposed CN_2. Thus we say for now that FR_3 is underspecified. The same holds for the requirement, correlated to CN_3 – it can be neither specified nor named at this stage and just shows once again why the axiomatic design postulates zigzagging among the domains during the design process. Similarly to FR_2, FR_4 is underspecified and can be determined/named only after working out the details at the lower levels. The requirement correlated to CN_4 could be formulated as *use of reliable modelling methods and tools*: FR_5.

Respectively, the requirement correlated to CN_5 is to *ensure model longevity*: FR_6. Let us summarize all these in a table (Table 4.1):

Table 4.1. Determining the *FRs* on the basis of *CNs*

#	Customer needs: perfect models...	Requirements: perfect modelling must...
0	...can resemble all important modellee traits	...allow representation of arbitrary attributes...
1	...can be used instead of their modellees	...ensure the model adequacy
2	...have the desired properties	...<underspecified>
3	...offer ease and comfort	...<underspecified>
4	...are robust and reliable	...use reliable modelling methods and tools
5	...live forever	...ensure model longevity

The numbering in the first column is used when referring to a row from the second or the third column. For compactness, ellipses indicate where a text is missing in front or at the end of a sentence. In this sense, demand 4 should be read as "perfect models are robust and reliable" (the column title and the 4th row jointly), demand 5, respectively, as "perfect models live forever" and so on.

Note that requirements 0 and 1 are specified, but their further elaboration is indispensable for the further analysis. Requirements 2 and 3 could not be specified at this stage because demands 2 and 3 are still not clear enough (underspecified). But how to define them and which additional requirements should be posed? In similar situations Suh suggests in his book about axiomatic design Suh (2001) to zigzag between the four domains (customer, functional, physical and process) in order to ensure optimal design. So it is appropriate to drive the iteration further and try to specify all ambiguous demands and requirements in more details.

4.2.1   First Approximation

Before doing the next iteration, let us consider the first attempt to formulate these requirements in detail: Avgoustinov (2004) defines six basic and ten auxiliary requirements for a "next generation" approach that would better satisfy the contemporary modelling demands. For easier reference they are cited below (with light improvements):

R1.   Maximally long lifespan of the models (*cf.* Section 3.1.1.8 above about the lifespan of different IT-components).

R2.   Global model access across all divisions of the enterprise.

R3.   Access to (or exchange of) parts of models.

R4.   Global (data/model) integration.

R5.   Use of product data management (PDM). Meanwhile this should be extended with use of Engineering Data Management (PDM) and Product Lifecycle Management (PLM).

R6.   (More) intuitive, natural and accurate modelling.

R7.   Capability to model product and process data (*i.e.* static and dynamic data).

R8. Distributed hosting based on platform-independence.

R9. Ability to keep the model-related knowledge within the respective models or objects (similarly to class properties and class methods by the OOP).

R10. High reusability of models and their parts.

R11. Capability to communicate offline (data exchange) and online (data sharing).

R12. Extensive support for cooperative work.

R13. Openness and extendibility: no matter whether it will be concept, architecture, approach, format or a combination of them, it should be and remain open and extendible.

R14. Little need for experts to act outside their field of expertise. This increases both the technical and the economical efficiencies.

R15. Either smooth transition of all legacy models and data to the new concept or their use in parallel must be guaranteed.

Of course, this list is far from complete. For instance, Westkämper (2000) mentions issues/topics of the "improved manufacturing", but only their first three groups (meeting user demands, quality improvement, and cost reduction) have been taken into consideration until now. The ever increasing demand for better and more detailed models, requiring also more computing power, suggests another requirement:

R16. Scalability of the complex (systems of) models.

This requirement could be satisfied by taking advantage of networks and building distributed systems. Ideally, therewith possibilities for cooperation would be extended too.

The modelling of (dynamic) processes by means of compound models provides an additional requirement, which can hardly be satisfied by models created in (different) CAx-systems: models with different origin, location and host should be able not only to communicate, but also to work with one another. This requirement – cf. Lee (2001) – or feature should be called:

R17. Interoperability of the created models.

This requirement also arises – although implicitly – from the combination of R4, R7, R8, R11 and R12.

The idea to employ the digital factory as a model of the (real) production process, e.g., as in Bley and Franke (2001), to achieve an optimal real product is nice, but leads to ever increasing complexity of the resulting model. After critical complexity is reached, neither keeping such a production model monolithic, nor keeping it homogeneous is possible. Thus, sooner or later the model becomes compound. Then the faster its complexity and number of components grow, the sooner the initially used CAx-system reaches its limits, and the necessity to prepare some of the sub-models with other CAx-systems arises. Since these systems are expensive, the price of complex compound models is also high. If the user and the developer of a model are different organizations, the necessity to acquire all the CAx-systems involved in the development can be unaffordable (especially for small enterprises) and thus turn into a real problem. To avoid this it is necessary that:

R18. Separated authoring and use of the models (also in independent environments) should be possible.

Some CAx-system developers have already recognized this need and offer the so-called viewers for free, in order to increase acceptance of their products. The viewers are thin applications capable of reading and visualizing models prepared with the CAx-systems from the same and, possibly, a few other brands. The main functional difference is that the viewers cannot modify the models. Since both the number of supported CAx-systems and the possibilities for integration of heterogeneous models are restricted, the use of viewers is not a (general) solution.

4.2.2   Demands, Requirements, Parameters

In addition to the (unordered) list of requirements, cited in Section 4.2.1 above, the problems and issues formulated in Chapter 3 have also to be considered.

For easier perception, they are classified and structured in several main categories and sub-categories numbered according to their estimated importance. The result is presented in Figure 4.1.

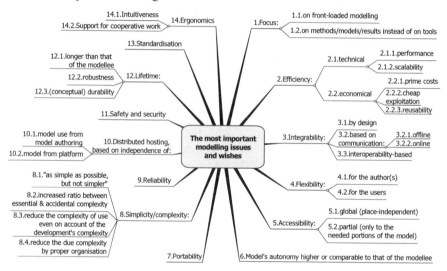

Figure 4.1. Main modelling issues

A brief comparison of the main modelling issues with the list of requirements from the previous section reveals that they have a lot in common. This is not surprising, since each requirement for a better approach pursues the elimination of one or more issues. Therefore, it is reasonable to compare them and see whether they complement each other.

4.2.3   Ordering the Requirements

Now we can return back to the formulation of the requirements for the perfect modelling approach. On the basis of the material presented in Sections 4.2.1 and 4.2.2 we can recognize, specify and partly prioritize the customer needs in more detail and define the respective requirements for the modelling down to their third sublevel. If necessary, they can be elaborated in even more detail later on. The first few levels of mapping of customer needs to requirements can look as presented in Table 4.2. Again, punctuation similar to that in Table 4.1 is used, but in addition, indentation denotes a lower level (if still unclear, the number in the first column should be used for orientation). Needs CN_0, CN_1, CN_2 and CN_3 are further decomposed.

Note that the word "delivery" is used in the sense of an unspecified way to obtain a model, which could be creation, modification/reuse of an available model, purchasing, or something else. The point is that the capability to create models quickly is important but creation from scratch is not the quickest way.

At this stage we could define FR_2 on the basis of its constituents as providing a *modelling (degree of) freedom* and FR_3 as *ensuring modelling expedience*.
The requirements formulated in Table 4.2 offer a reasonable basis for the next step – finding appropriate *modelling parameters* (since we are discussing modelling, this term will be used instead of design parameters, as more appropriate), which would allow us to satisfy the requirements proposed until now. Before determining and specifying these parameters, though, it would be appropriate to review some other attempts to develop a better approach than the SCA, and try to avoid their disadvantages, while making best use of their achievements and benefiting from their ideas and experience.

Table 4.2. Determining the *FRs* on the basis of *CNs*, level two

#	Customer needs: perfect models...	Functional requirements
0	...resemble the modellees' most important traits, including...	...allow representation of arbitrary attributes...
0.1	...outlook	...outlook
0.2	...functionality	...functionality
0.3	...behaviour	...behaviour
1	...can be used instead of their modellees...	ensure the model adequacy:
1.1	...and lead to comparable results	ensure model plausibility
1.2	...for earlier and faster decision making	ensure quick delivery
1.3	...for cheaper decision making	ensure low running costs
2	...can...	provide modelling freedom:
2.1	... represent products and processes	allow static and dynamic modelling
2.2	...have any type of aspects	design for extensibility
2.3	...have any number of aspects	unrestricted model size
2.4	...have any size	ensure scalability
2.5	...be of any kind (domain-independent?)	use of generic modelling techniques
3	...can be easily and effortlessly...	ensure modelling expedience:

Table 4.2. Determining the *FRs* on the basis of *CNs*, level two (continued)

#	Customer needs: perfect models...	Functional requirements
3.1	...obtained...	provide efficient modelling concept(s):
3.1.1	...with minimal effort	minimal number of instructions for the accomplishment of any task
3.1.2	...with minimal costs	ensure cheap creation and reusability
3.1.3	...in minimal time	ensure quick delivery
3.1.4	...without dedicated qualification	minimise the accidental/due complexity
3.2	...integrated with other models	design models ready for integration
3.3	...used	keep models simple and intuitive
3.4	...adapted to newly emerged needs	flexibility on all hierarchy levels
4	...are robust and reliable	use of reliable modelling methods and tools
5	...live forever	ensure model longevity

4.3   Attempts to Avoid the Drawbacks of Conventional Modelling

4.3.1   Dissatisfaction of Modellers with Existing Solutions

CAx-systems have gone through a long development with countless improvements, extensions and enhancements. The organization, control and most of all the automation of production (design, manufacturing, *etc.*) are nowadays unthinkable without them. Some systems have already such extensive functionality that many people think of them as all-mighty. On the other hand, it seems that users are not fully satisfied with SCA: each couple of years a new CAx-system type arises with the ambition to bring along the missing features as well as processing (or modelling) capabilities that will cover the unsatisfied user demands. Researchers seem to be unsatisfied too, since numerous novel alternative approaches, methods and tools, trying to avoid disadvantages of SCA, are permanently developed and introduced. Some of them attempt to extend SCA-created models with new functionality, other try to achieve the same or better results by alternative means. The number of these efforts not only confirms the existence of flaws in SCA, but also speaks of their severity.

Some more important examples are discussed below, whereas distinctive formatting style is applied to more important disadvantages: they start by a prefixed with "Disadvantage" number for easier reference later on.

4.3.2 Cooperative Modelling with Transient Objects (Integrated Engineering System)

The *cooperative modelling with transient objects* approach (also known as *integrated engineering system* Gausemeier *et al.* (1996)) is based on an object-oriented system design and a CORBA-based access to the internal models of the used CAE-tools. It introduces three levels of integration – system, model and process – that correspond to layers 1–5, 6 (representation layer) and 7 (application layer) of the ISO/OSI reference model, respectively. The integration is achieved by means of an application programming interface (API), remote procedure calls and inter-object communication at the system level; by file-oriented data exchange and a common product model at the model level; by means of a dedicated process management module at the process level. The CAE-tools are encapsulated intelligent objects, communicating with an object request broker (ORB) at the system level, and using STEP-based representation at the model level. As stated in Gausemeier *et al.* (1996, p. 323), the transience of these objects allows "higher dynamic" of the system. A disadvantage is:

Disadvantage 1. Many CAx-developers abstain not only from adopting STEP as internal format but also from incorporating STEP converters in their systems (due to their size and complexity). Since the "intelligent object" is on average smaller than a CAx-model, choosing a STEP-based representation for them can be inefficient or suboptimal.

4.3.3 Network-centric Virtual Prototyping

The main idea of the *network-centric virtual prototyping* project is to provide a global network-centric and spatially distributed environment, "which enables product designers to communicate more effectively, obtain and exchange a wide range of design resources during product development" – *cf.* Lee (2001). The approach builds upon a CORBA-supported client–server architecture with processing distributed among the server and the clients. The approach relies on geometry and other model data coming from CAx-systems over STEP or other neutral formats. This approach only pertains to a sub-area of the product and process modelling, but is mentioned here since it confirms the doubts about the adequacy of the system-centred approach.

4.3.4 Component-based CAx-systems

The *component CAx-systems* approach is presented and discussed in numerous publications, *e.g.*, in "ANIKA"), Dankwort (1997), Kilb and Arnold (1998), Arnold *et al.* (1999). According to Poppendieck (2004), its roots lie more than 200 years back in history:

Around 1800, Eli Whitney proposed manufacturing rifles with interchangeable parts, instead of crafting each rifle individually. Widely regarded at the beginning of mass production, the concept of interchangeable parts led to a dramatic increase in rifle production capacity while delivering the additional benefits of consistent operation and easy field maintenance of weapons.

Component-based systems represent a paradigm shift in software development similar to that of using interchangeable parts in manufacturing. A component-based system is built of standard, reusable parts that become the fundamental building blocks of future software. Component-based systems promise numerous benefits, including flexibility, scalability, and maintainability.

Although he speaks about software systems in general, this is true for CAx-systems, too. In short, the idea is to develop component based CA-systems, which have more modern and adequate architecture and gradually replace the conventional CAx-systems. According to the systematic review of the component-based technologies and their application in the field of CAx-systems made in Janocha and Gandyra (1997), it is essential for a CAx-component to be a binary, clearly identifiable building block with well-defined, **STEP-based interface** which implements the relevant CAx-functionality for a given subject area. The integration of models, created in different CAx-systems, is arranged online by a so-called "CAx object bus". Conventional CAx-systems are supported by the so-called ANICA-adapters (ANICA is abbreviated from **AN**alysis of access **I**nterfaces of various **CA**x-systems). A disadvantage is:

Disadvantage 2.   The user of any compound model ought to possess (or to have access to) all involved in the modelling CAx-systems. R11 is not fully satisfied. For each CAx-system involved, separate computer (two systems cannot run simultaneously on the same computer) and connections to the others are required. It is expected that more developers would be able to supply CAx-components than CAx-systems, since the former are smaller and easier to develop, but Disadvantage 1 holds here, too.

4.3.5   Product Data Markup Language (PDML)

The *product data markup language (PDML)* was proposed by William Burkett in 1998 as an "application of STEP technology" based on database scheme modelling principles – *cf.* Burkett (1998). It aims at some improvements – for instance to leverage the abstraction and context-sensitivity of the represented information. The most interesting trait of PDML is the idea to rely on a generally available, Web-friendly, universal representation language: XML (eXtensible Mark-up Language). A disadvantage is:

Disadvantage 3.   PDML has no native means for representing (lower level) algorithms and consequently does not comply (at least) with R7, R11, R12, R15 and R17.

4.3.6   PDM Enablers, PDM Schemas

An overview of the *PDM enablers* and the *PDM schema* is given in Starzyk *et al.* (1999). According to Goltz (2000) the PDM systems offer meanwhile good capabilities to manage the amount of data, but only for a single company and not for a distributed engineering environment. One of the known efforts to overcome these drawbacks by means of *federated PDM and database systems* is described in Abramovici *et al.* (1998), but there is a danger – that they could lead to a situation like the one that initiated the development of STEP. Disadvantages of the PDM enablers are at least that they are:

Disadvantage 4.   Well-suited to only a single company.

Disadvantage 5.   Not suitable for distributed engineering environments.

Disadvantage 6.   Unable to offer solutions for all problems and issues discussed in the previous chapter.

4.3.7   Innovative Technologies and Systems for Integrated Virtual Product Creation (iViP)

The project "Innovative Technologies and Systems for the integrated Virtual Product Creation" – also known as iViP (*cf.* Krause *et al.* (2002)) is supported by the Federal Ministry of Education, Science, Research and Technology and unites 51 partners from industry and research. It started in 1998 for a period of four years and aimed at preparing technology and tools for the heterogeneous and completely integrated process flow of tomorrow. The iViP has CORBA–based, server–client modular architecture, providing the basis for an open integration platform and supporting also existing heterogeneous system worlds. The architecture seems to have some similarity with the CAx object bus of the project "ANIKA" – *cf.* "ANIKA" (1998), but offers additional tools to the user. Although at the time of writing this overview iViP is not yet finished and the available information is scarce, the approach is very promising.

4.3.8   Process-centred Development

The component-based approach *process-centred development* is presented in Jesko and Endig (2000) with a relatively simple UML-model with seven (main) components.

4.3.9   PDGL

Part Design Graph Language (PDGL), described in Krause *et al.* (1991) and VDI 2218 (p.46) is a formal language for description of features, developed at TU-Berlin and based on the standardized in ISO 10303 language EXPRESS. The main disadvantage of PDGL is the fact that it inherits a lot of complexity from EXPRESS. A standard implementation of PDGL – if it exists – does not seem to be very popular.

4.3.10   Model Driven Architecture (MDA)

MDA was developed by the Object Management Group (OMG) and adopted in September 2001. It conforms with the OMG's mission: to develop an architecture for distributed application integration, using object-oriented technology and guaranteeing reusability of components, interoperability, portability and (common) basis in commercially available software, with freely available specification - Soley (2002), Soley (2005). According to Poole (2001), it is based on and extends the important OMG standards The Unified Modelling Language (UML), Meta Object Facility (MOF), XML Metadata Interchange (XMI), and the Common Warehouse Metamodel (CWM). These standards define the core infrastructure of the MDA, and have greatly contributed to the current state-of-the-art of systems modelling.

According to Frankel (2001, p.31) modelling languages can be used as programming languages to improve productivity, quality and longevity.

MDA offers excellent concept and architecture for implementing both authoring tools and systems of models. It considers many of the information-technology-related issues of the modelling, but on a relatively low-level of implementation. Therefore, although very suitable for implementation of software models, it seems to be unable to solve all problems and issues of mechanical and mechatronical modelling – at least not without extensions and improvements.

4.3.11 Holonic Manufacturing

Let us first introduce some definitions for clarity:

Holon: object, which can be viewed as both a part of a given system and as an autonomous or compound entity.

Manufacturing Holon: An autonomous and cooperative building block of a manufacturing system for transforming, transporting, storing and/or validating information and physical objects. The manufacturing holon has always an information processing part and often a physical processing part. A holon can be part of another holon.

Autonomy: The capability of an entity to create and control the execution of its own plans and/or strategies.

Cooperation: A process whereby a set of entities develops mutually acceptable plans and executes these plans.

Holarchy: A system of holons that can cooperate to achieve a goal or objective. The holarchy defines the basic rules for cooperation of the holons and thereby limits their autonomy.

Holonic Manufacturing System (HMS): A holarchy that integrates the entire range of manufacturing activities from order booking through design, production, and marketing in order to realize the agile manufacturing enterprise.

Holonic Attributes: The attributes of an entity that make it a holon. The minimum set is autonomy and cooperativeness.

Holonomy: The extent to which an entity exhibits holonic attributes.

Some typical examples of manufacturing holons are: continuous processing holon, machining holon, assembly holon, transportation holon and system optimization holon.

In comparison to other well-known notions a holon is cooperative like a server in a client–server environment and autonomous like a free agent (http://www.mech.kuleuven.be/goa/). To summarize, the holonic approach puts the focus on decentralization of the control and decision making, which is achieved by incorporation of intelligence in the holons. Sometimes advantages specific to the fractal manufacturing (see below) can be also observed.

4.3.12 Fractal Manufacturing

According to Warnecke (1996) the term *fractal* originates from the Latin word *fractus* (broken, fragmented). It is usually described in the literature as a *self-similar* geometry or structure, but this definition is somewhat unintuitive and inaccurate. What is actually meant by self-similar is that there is some pattern –

either structural, or geometrical, or of whatever nature – which can be found or recognized on different hierarchy levels within a given object or system.

Fractal Factory: self-similar, self-organized, self-optimizing, goal-oriented, dynamic structure, *cf.* Warnecke (1996). Goals are also self-similar.

A (manufacturing) fractal can be viewed as a special kind of holon: it is by definition part of a bigger (or higher level) fractal and can be simultaneously independent and self-functioning. But on another level in the hierarchy there always exists a fractal with similar organization, topology, structure, ways of control, *etc.* These similarities can lead to many advantages – mainly reuse of work and simplification. The most important advantage of the approach is that identical (control) structures, used on the different fractal levels, lead to more intuitiveness and easier predictability, simplifying therewith the control. By means of software systems reusability of the control modules is achieved for similar fractals (components with the same properties/structure), which leads to cost reduction and increases the efficiency.

4.3.13 Others

The above survey of some known approaches does not pretend to be complete or exhaustive: there exist some approaches that are known but not mentioned here because they are not (fully) relevant to our presentation scheme. Other approaches might be relevant, but still unknown to the author. Some other approaches are really worth mentioning, but are not included in the survey due to time-related reasons. Among them are:

- Bionic Manufacturing
- Property Driven Design (PDD); KFT, Saarland University described in Weber *et al.* (2003; Weber (2005a;b)
- Global Manufacturing Interface (GMI), University of Twente
- Reference Model of Open Distributed Computing: ISO/IEC 10746
- Modular Factory
- Multi-agent Systems (Bochmann *et al.* (2003; Fischer *et al.* (2003), *etc.*)
- BEA WebLogic Integration (BEA Systems (2004)
- Service Oriented Architecture (SOA)
- ESPRIT 21955: Working Group on Intelligent Manufacturing Systems (http://www.mech.kuleuven.be/imswg/welcome.html)
- Intelligent Mechatronic Systems (http://www.mech.kuleuven.be/imechs/ or http://www.mech.kuleuven.be/ams/)

4.4 Modelling Parameters

After detailing so many and different requirements, we have to consider what means can be used to satisfy them. Since the focus of this book is more on methodology than on software implementation we shall neither go into too much implementational detail, nor consider specific solutions or examples.

4.4.1 Parameters Enumeration

At the beginning, let us just review which quantities, attributes or traits of the models and of the modelling could be influenced directly: those are our modelling parameters. A handful of parameters well-known in software design are given in Figure 4.2 together with modelling parameters, which I know from my personal practical experience. They are grouped in three main categories: strategic/conceptual activities, model organization and implementation modalities. Of course, this grouping is neither absolute nor final, but it helps in the understanding and handling of the matter.

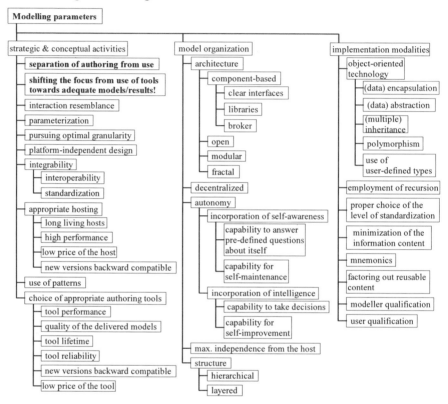

Figure 4.2. Most important modelling parameters

As its name suggests, the requirements in the strategic/conceptual group seem to be the most important or have greatest impact on satisfaction of the requirement. This could be proved by considering their relations to the main (functional) requirements.

4.4.2   Main Relations Between (Modelling) Parameters and (Functional) Requirements

The functional requirements (*FRs*) detailed in 4.2 and modelling parameters from Figure 4.2 (*DPs*; for now only the strategic ones) are listed side by side in Table 4.3. The meaning of ellipses is the same as in Table 4.1. Note that no horizontal lines are drawn to stress that no correspondence or mapping is established until now.

Table 4.3. Determining the dependence of *FRs* on the *DPs*

Functional requirements	Modelling parameters
perfect modelling must...	separation of authoring and use
...allow representation of arbitrary attributes...	focus on achieving adequate results
...outlook	
...functionality	interaction resemblance
...behaviour	parameterization
...ensure models' adequateness...	pursuing optimal granularity
...ensure model plausibility	platform-independent design
...ensure quick delivery	integrability
...ensure low (running) costs	interoperability
...provide modelling freedom:	standardization
allow static and dynamic modelling	use of patterns
design for extensibility	host...
unrestricted model size	...long living hosts
ensure scalability	...host's performance
use of generic modelling techniques	...low price of the host
...ensure modelling expedience:	new versions backward compatible
provide efficient modelling concept(s)	choice of appropriate tools
can accomplish any task with minimal number of instructions	...performance
ensure cheap creation and reusability	...quality of the delivered models
...ensure quick delivery	...lifetime
minimize the accidental/due complexity	...reliability
design models ready for integration	...backward compatible
keep models simple and intuitive	...low price of the tool
flexibility on all hierarchy levels	
use reliable modelling methods and tools	
ensure long model-lifetime	

The next step towards a better (solution for) modelling is to find the interdependencies among the functional requirements and the modelling parameters. More precisely, in order to satisfy all requirements it is necessary to know which of them depend on which parameter, and if there are parameters that impact more than one requirement their exact influence should be known. Since the determination of these interdependencies – in our case about twenty requirements and about forty-five modelling parameters – takes plenty of time, only the results of this investigation are given here, followed by some comments. Note that these results can differ if the same analysis is performed from a team with different experience, as well as if the sets of requirements and parameters are changed due to differences in the analysed case.

The easiest possibility to represent these results is to use a matrix, similar to the design matrix used in the axiomatic design Suh (2001) or in the property driven development/design (PDD) Weber *et al.* (2003); Weber (2005a;b). This is illustrated in Table 4.4 and shows how the existence (or the change in the value of) a parameter influences the satisfaction of one or more functional requirements. A "x" in a cell of Table 4.4 denotes an apparent dependence in direct proportion to the requirement, written as heading of the respective column, on the parameter, written as heading of the respective row. A "*x-1*" denotes, respectively, inversely proportional influence. A question mark means that the dependence is probable, but not sure. Dependencies needing further investigation are denoted "*n.i.*" (abbreviated from "need investigation").

If you are familiar with the *quality function deployment* (QFD) technique, Table 4.4 could seem familiar to you or you could even ask yourself, is it possible to use QFD here instead or at least to estimate more precisely each relation: yes, it is possible. Initially we wanted to give estimation for the strenght of each relation in Table 4.4, but we decided that when working with so many requirements and parameters it is too difficult to assess the relation strengts properly and consistently troughout the whole table when viewing a general case like the case here.

4.4.3   Rank of Influence

The next step should be to consider the importance of each parameter, or its rank of influence on satisfying the functional requirements. This can be performed on the basis of Table 4.4, which illustrates the interrelations between the requirements and the parameters. Since both of them are actually represented as hierarchical trees, only the leafs of these trees should be compared, otherwise their parents (or superordinates) could show mixed relations, which can be misleading.

In Table 4.4 (page 168) there is one row (labelled "degree of dependence") and one column (labelled "rank of influence"), performing special tasks: they show the count of the related items for the corresponding category. In other words, the degree of dependence of any requirement shows the number of parameters it depends on; the rank of influence shows the number of requirements depending on a given parameter. Although these measures are only quantitative, they can give first impressions about which parameters are more influential and which requirements can be critical to achieve due to a too high degree of influence. Both measures consider only the directly proportional influence. Since no qualitative measures can be determined for now, there is not much choice but to investigate the quantitative aspect.

Let us begin with the strategic and conceptual parameters. A brief look at the table reveals that 5 of the 19 parameters have influence on half of the 20 requirements or more, whereas 4 requirements depend on the half of the strategic/conceptual parameters. As can be seen, the highest rank of influence – 13 – is assigned to *interoperability*, followed by *parameterization*, *platform-independent design* and *use of patterns*, with rank of 12 each. Thereafter follow *shifting the focus* (on the models instead of tools) with 10, *separation of authoring from use* with 9, and *quality of the provided models* also with 9. A short discussion of these parameters is provided in Section 4.6.

Note that when a specific solution is considered, achieving cheap creation and low running costs are usually viewed as constraints and not as functional requirements. But since here they are the result of the search for a generic solution (or approach), they are considered as requirements in this case and have to be analysed. Also noteworthy is the fact that the prices of the authoring tool and the host seemingly have the lowest rank of influence – 1. Actually, despite these low ranks the real influence of these two parameters is considerable and will be discussed later on.

4.5   Model Centred Approach

> *I have had my results for a long time: but I do*
> *not yet know how I am to arrive at them.*
>
> Karl Friedrich Gauss

A new – not burdened with obsolete restrictions, applications and standards – approach could allow recent achievements in computer science to bring "fresh air" to product and process modelling. But is it at all possible to satisfy the requirements presented in Section 4.2 above, if the powerful CAx-systems cannot? If Nature is considered for a moment, numerous examples illustrating the (power of) natural selection can be found: bigger is not always stronger, groups of small animals often achieve more than one big animal; smaller creatures often adapt better and propagate quicker than the bigger ones.

4.5.1   Idea

The careful consideration and assessment of the techniques presented in Section 4.3 on the basis of the available information, among other reasons, have led us to conceive an approach to complement the SCA, as well as to provide an alternative possibility for development of sophisticated, but less complex systems of models. This approach – the *Model Centred Approach* (or in short MCA) – aims at avoiding or eliminating the problems and disadvantages mentioned in the previous chapter. Its initial description has been presented in Avgoustinov (2002), and Avgoustinov (2004). Driving force of the approach is the modelling, complemented by concepts like intuitiveness, object-oriented and feature-based design, separation of model authoring and use, multiple levels of detail, aspects, *etc.*

Although the SCA has been historically predetermined, it is not relevant for contemporary conditions and cannot deal anymore with its own deficiencies and with the problems arising. It seems much more appropriate and relevant if product and process modelling is performed on an aspect- and object-oriented, model centred basis, where the main building blocks are components, implemented as autonomous intelligent entities – the main driving force. The "name" model centred approach has been chosen as opposite to "system centred approach" and to stress the fact that the priorities of the MCA are different.

Table 4.4a. Modelling parameter influence on functional requirements (UL part)

				...outlook	...functionality	...behaviour	...ensure model plausibility	...ensure quick delivery	Correlation number
degree of dependence				3	10	12	6	5	dd
				0.1	0.2	0.3	1.1	1.2	
			separation of authoring from use				x		1
			shifting the focus from use of tools towards adequate models/results!		x	x	x	x	2
			interaction resemblance		x	x	x		3
			parameterization	x	x	x		x	4
			pursuing optimal granularity						5
			platform-independent model design		x	x		x	6
		integra-bility	interoperability		x	x			7
			standardization		x	x			8
			long living hosts		x	x			9
		approp-riate hosting	high performance			x	x		10
			low price of the host						11
			new versions backward compatible		x	x			12
			use of patterns		x	x		x	13
			tool performance					x	14
	choice of appropriate authoring tools		quality of the delivered models	x		x	x		15
			tool lifetime				x		16
			tool reliability						17
			new versions backward compatible	x	x	x			18
			low price of the tool						19

Modelling parameters — strategic & conceptual activities

Correlation number 1| 2| 3| 4| 5| +

Table 4.4b. Modelling parameter influence on functional requirements (UR part)

Correlation number	...ensure low (running) costs	allow static and dynamic modelling	design for extensibility	unrestricted model size	ensure scalability	use of generic modelling techniques	can accomplish any task with minimal number of instructions	ensure cheap creation and reusability	minimize the accidental/due complexity	design models ready for integration	keep models simple and intuitive	flexibility on all hierarchy levels	use of reliable modelling methods and tools	ensure long model-lifetime	rank of influence
dd	7	6	5	5	4	5	2	11	5	6	8	6	4	10	
	1.3	2.1	2.2	2.3	2.4	2.5	3.1.1	3.1.2	3.1.4	3.2	3.3	3.4	4	5	
1	X			X	X			X	X		X	X		X	9
2								X	X	X			X	X	10
3		X				X		X		X					7
4			X		X	X		X		X		X		X	12
5	X			X	X				X	X	X	X		X	7
6		X	X			X		X	X	X	X			X	12
7	X	X	X		X	X		X	X	X	X	X		X	13
8	?x^{-1}		x^{-1}	n.i.	?			X	x^{-1}	X		X		X	6
9	X							X						X	5
10	X	X		X											5
11	X														1
12	X							X						X	5
13		X	X			X	X	X			X	X	X		12
14		X		X											4
15			X	X		X					X		X	X.	9
16								X							2
17												X			1
18								X							4
19								X							1

+ | 6| 7| 8| 9|10|11| 12|13|15|16|17|18|19|20| R| |

Table 4.4c. Modelling parameter influence on functional requirements (ML part)

					...outlook	...functionality	...behaviour	...ensure model plausibility	...ensure quick delivery	Correlation number
degree of dependence					1	10	8	3	5	
					0.1	0.2	0.3	1.1	1.2	
Modelling parameters	model organization	architecture	compo-nent-based	clear interfaces		x	x		x	20
				libraries		x	x		x	21
				broker					x	22
				open					x	23
				modular						24
				fractal						25
			decentralized		x	x			26
		autonomy and intelligence	incorpo-ration of self-aware-ness	capability to answer pre-defined questions about itself		x	x			27
				capability for self-maintenance		x	x			28
			incorpo-ration of intelli-gence	capability to take decisions		x	x			29
				capability for self-improvement		x	x			30
		max. independence from the host			x	x		x	31
		structure	hierarchical	x					32
			layered	x					33

Correlation number | 1| 2| 3| 4| 5| +

Table 4.4d. Modelling parameter influence on functional requirements (MR part)

Correlation number	...ensure low (running) costs	allow static and dynamic modelling	design for extensibility	unrestricted model size	ensure scalability	use of generic modelling techniques	can accomplish any task with minimal number of instructions	ensure cheap creation and reusability	minimize the accidental/due complexity	design models ready for integration	keep models simple and intuitive	flexibility on all hierarchy levels	use of reliable modelling methods and tools	ensure long model-lifetime	rank of influence
	2	1	5	1	3	9	5	10	6	4	9	3	8	4	
	1.3	2.1	2.2	2.3	2.4	2.5	3.1.1	3.1.2	3.1.4	3.2	3.3	3.4	4	5	
20	x		x			x		x		x	x	x		x	12
21	x		x			x		x				x		x	10
22	n.i.		x												3
23	x	x	x					x		x		x		x	9
24	x	x	x			x		x		x	x	x	x		9
25	x							x	x		x				4
26				x	x	x								x	6
27			x					x		x				x	6
28	x				x					x				x	6
29	x				x					x	x				6
30	x				x					x	x			x	7
31	x		x	x	x			x	x		x			x	12
32		x	x	x	x	x			x	x	x			x	10
33		x	x			x			x	x	x	x	x	x	10

+ | 6| 7| 8| 9|10|11| 12|13|15|16|17|18|19|20| R| |

Table 4.4e. Modelling parameter influence on functional requirements (LL part)

				...outlook	...functionality	...behaviour	...ensure model plausibility	...ensure quick delivery	Correlation number
				6	28	28	9	15	
				0.1	0.2	0.3	1.1	1.2	
Modelling parameters	implementation modalities	object-oriented techno-logy	(data) encapsulation		x				34
			(data) abstraction		x				35
			(multiple) inheritance		x	x		x	36
			polymorphism		x	x		x	37
			allow user-defined types				x		38
		employment of recursion			x	x			39
		proper choice of the level of standardization			x	x		x	40
		minimization of the information content			x	x		x	41
		mnemonics			x	x	x		42
		factoring out reusable content			x	x		x	43
		modeller qualification			x	x	x		44
		user qualification		x					45

Legend:

x: the requirement in the column depends on the parameter in the row

x^{-1}: the same as above, but the proportion is inverse

?: uncertainty of the dependence

Correlation number: a number that allows correlation of rows or columns of parts of the table, appearing on different pages

UL=upper left part of the table

UR=upper right part of the table

ML=middle left part of the table

MR=middle right part of the table

LL=lower left part of the table

LR=lower right part of the table

Table 4.4f. Modelling parameter influence on functional requirements (LR part)

Correlation number	...ensure low (running) costs	allow static and dynamic modelling	design for extensibility	unrestricted model size	ensure scalability	use of generic modelling techniques	can accomplish any task with minimal number of instructions	ensure cheap creation and reusability	minimize the accidental/due complexity	design models ready for integration	keep models simple and intuitive	flexibility on all hierarchy levels	use of reliable modelling methods and tools	ensure long model-lifetime	rank of influence
	18	11	19	9	13	20	7	28	18	18	23	14	14	24	
	1.3	2.1	2.2	2.3	2.4	2.5	3.1.1	3.1.2	3.1.4	3.2	3.3	3.4	4	5	
34						x	x	x		x			x	x	7
35						x	x	x		x			x		6
36			x			x	x	x					x		9
37		x				x	x	x		x		x	x		11
38		x	x			x	x	x	x	x	x	x	x		11
39						x	x				x		x		6
40	x		x	x	x			x	x	x	x	x		x	14
41	x					x	x		x		x				9
42						x		x	x		x			x	8
43					x	x		x			x		x		9
44			x					x		x	x		x	x	9
45						x	x								3

4.5.2   About the Name

> *What's in a name? That which we call a rose*
> *By any other name would smell as sweet.*
>
> William Shakespeare
> Romeo and Juliet. ACT II Scene 2.

Two approaches, which became very popular during the last decade, are the *model-driven architecture* (MDA), initially described in Poole (2001), which is a registered trademark of OMG[40], and Model-Driven Engineering (or MDE). According to Wiki (2006), MDE "*...refers to the systematic use of models as primary engineering artefacts throughout the engineering lifecycle. MDE can be applied to software, system, and data engineering*", and its best initiative is the MDA.

A reasonable question arises regarding whether MCA is really different from MDA/MDE, and if yes, where the difference is. Since we not only agree with the principles of MDE and MDA, which we are aware of, but also follow most of them, MCA has a lot in common with these approaches. Nevertheless, there are differences – or additions – in MCA, which we want to emphasize. One of the ways to do this is to start with the name of the approach. Since every name is a model of the respective thing, it has to reflect its most important traits. The terms MDA and MDE leave us with the impression that the model is somehow available and from now on it (or its use) will drive the engineering (or respectively, the architecture). In fact, they are much more concerned with the proper and efficient implementation of the models, whereas the MCA deals more with the model emergence and model use.

In (mechanical) engineering it holds that the earlier a certain error is made in multi-stage development and the later it is discovered, the worse its negative consequences are. For this reason there is a trend to so-called front-loaded development, aiming at achieving better overall efficiency of the development by means of investing more effort in the early discovery of each possible error. Since every product or production development starts with a model, we shall be more concerned with the model emergence, because a proper development would lead automatically to better use. Therefore, we have tried to choose a name stressing the fact that every model (with its properties) is much more important than the tools used to prepare it or even the expected product.

4.5.3   Definitions

4.5.3.1   Authoring and Authoring Tools

We shall refer to the process of creating artefacts in its most general sense as *authoring*. Modelling is a kind of authoring. Production or manufacturing are, in contrast, not necessarily bound to the authoring of a product, since the product to be manufactured emerges during the design (*i.e.*, much earlier) and only the

---

[40] Object Management Group (OMG) has been an international, open membership, not-for-profit computer industry consortium since 1989. A list of its registered trade marks is available on its web-page, http://www.omg.org/legal/tm_list.htm.

replicas of a product appear as result of production or manufacturing. The latter is itself a result of authoring, and the respective process is called production planning.

An author can use one or more *authoring tools* for his (main) activity. Since this term reflects only the function of the tool and not its type, it will be used here for referring to extremely different types of software tools – from a simple text editor to a sophisticated CAx-system or a dedicated modelling environment – as long as the context concerns (model) creation.

4.5.3.2 Host

The created models have to live somewhere – we named the respective environment *host*. Strictly speaking, the host consists of two main components – (computer) hardware and software – whereas in the optimal case the software component is infinitely small or non-existent. In our further discussion we concentrate our attention on the software and under host we mean only its software component, if not explicitly otherwise specified.

The host software component can have one or several layers. Typically, the lowest layer is the so-called *basic input–output system* (a.k.a. BIOS), the second layer is the *operating system* (or OS), and the third layer is a *dedicated software*. Since the first two layers are generic and their functionality is not used entirely, in some special cases – *e.g.*, for use in embedded systems – the dedicated software implements only the needed functionality of one or both lower layers. As a consequence the resulting software component of the host becomes leaner, needs less resources and therefore becomes more efficient.

4.5.3.3 Synthesis

There seem to be two ways in which a model or a solution to a given problem can emerge: either through gradual development or through testing different combinations of existing sub-models (or sub-solutions), and composing from them, the sought one. The latter way is known as synthesis (*cf.* Figure 3.18) and is much more widespread than one would initially think. For instance, we go through a similar process each time we want to speak: what really happens is that first we seek the necessary notions (the components), then the corresponding terms (the representation of the components), and finally all these words are put together in a phrase – the result of the synthesis. Recalling how often we (have to) speak, it would not be exaggerated to say that our brain is much better trained to synthesize than to perform a gradual development from scratch.

Applied to modelling, this means that it could be much more efficient to build up large/complex models by means of synthesis than by means of gradual development from scratch. Consequently, the componentization of the models, followed by a synthesis-based development of the models at the upper hierarchy levels, is one of the most important steps towards a better modelling approach.

4.6   Inherencies of MCA

4.6.1   A Meta-solution

The MCA is a collection of methods for product and process modelling, which have to support the models throughout their full lifecycle and especially throughout the development phase[41] of this lifecycle. The MCA is not bound to (the use of) specific tools, but recommends for some activities, specific types of tools; for this reason we call it a *meta-solution*. This meta-solution aims at better organisations of the models in compound models or in systems of models, but, of cource, it requires some technical and conceptual improvements, too.

The easiest way to describe MCA is to enumerate the most important of the methods and concepts involved – the MCA commandments.

4.6.2   The MCA-commandments

1. Separate (the tools for) model authoring from (the tools for) model use
2. Shift the focus from use of tools towards pursuing adequate models/results
3. Model for reuse
4. Make the models end-user friendly (not author/designer friendly)
5. Take advantage of autonomy and intelligence
6. Recognize patterns and features and employ them
7. Try to make models platform-independent and even host-independent
8. Use appropriate hosting
9. Pursue optimal granularity

Some of these commandments are briefly explained below. Since the activities, related to (the achievement of) some of them are interwoven, it is not easy to draw clear boundaries and to fully avoid repetitions.

4.6.3   Separation of Authoring from Use

The separation of the creation of models from their use[42] has the greatest impact on the modelling process – with almost an avalanche effect on efficiency, price, speed, *etc*. The idea behind this is that the authoring and the use of software models are two different phases of the model lifecycle. These phases have different timespans – ideally, a model would be prepared for a couple of seconds, but would be usable forever – and imply very different functionality. Consequently, it is logical and reasonable to use separate dedicated applications to support of each of these phases instead of trying to use one CAx-system for the support of both

---

[41] A good description of this phase for mechanical products, together with recommendations on how to put it into practice, is given in the guidelines number 2221 VDI-Richtlinien (1993) of the Union of German Engineers (VDI).

[42] Such separation is natural to all real (as opposed to virtual) artefacts, but as mentioned in Avgoustinov and Bley (2006), for historical reasons not (yet) applicable to most software models.

phases. These separate applications (or software systems), named authoring tools and hosts in Section 4.5.3 above, can be developed in such a way that the hosts implement *only the necessary* functionality for the respective phase. So, every host can remain much simpler and more compact than any authoring tool, and can be viewed as a kind of lean software system. This, in turn, has multiple effects and leads immediately to several advantages, some of which are listed below:

- reduction of the accidental complexity (Brooks Jr. (1987) – since the hosts are lean, they are much simpler and easier to deal with than an average CAx-system;
- shifting the focus of the effort onto the result, rather than the tool – the user of a model does not have to learn sophisticated CAx-systems;
- getting the most out of experts due to proper focusing and more efficient distribution of the effort – each employee can concentrate on the matter he is expert in;
- getting the most out of hardware which hosts the models, due to more efficient distribution of the resources.

Such a separation may seem to have at least one unsolvable problem: the integration of the components of compound models. And it may seem unsolvable if one considers it from the viewpoint of conventional modelling – integration of models, which are created by means of heterogeneous systems *and* live in these systems.

The question here is not whether it is possible – a prototype implementation has been used to modell different products and processes in engineering areas like assembly Bley *et al.* (2002), machining simulation Avgoustinov (2000a), education Avgoustinov (2000b), layout planning Avgoustinov and Bley (2003) and others. Instead, the question is about the best way to achieve this separation, and eventually the best way to achieve the integration of separately created models.

One of the best possibilities is to prepare the models as autonomous components by incorporating intelligence and communication capabilities into them. Thereafter these models can be stored into a (public) repository, which could be viewed also as a "component market", and complemented with a broker as in the CORBA concept – *e.g.*, in Siegel (2000). Such an approach can be followed even by existing CAx-systems, allowing them to behave conform to MCA. A description of a test-bed application is presented in Bley *et al.*) and illustrated in Figure 4.3.

4.6.4 Separate Modelling

With the increase in size and complexity of a given mechatronical model the need for more computing resources (processor power, memory, visualization capabilities, *etc.*) also increases. If the model cannot be improved anymore to manage (or to live) with less resources, two main choices remain: either to involve even more resources – a "brute force" method – or to reduce the number of other programs that are running on the same computer in parallel but are not really needed.

Ironically, the number of the modellers (or sometimes – managers) tempted to apply the "brute force" method seems to be much higher than the number of those trying to find alternative methods. A probable explanation is that it seems very easily and quickly achievable, especially if the so-called "plug and play" method is

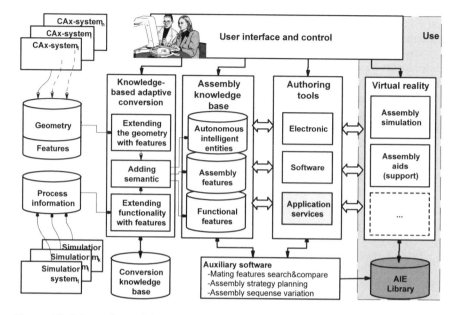

Figure 4.3. Scheme for applying MCA to existing conventional models after Bley *et al.*)

applicable: acquire additional hardware – there is usually a wide selection available – install it and go on. Sometimes, though, the scalability of this approach soon comes to its boundaries due to technical or organizational reasons.

In computer science the problem has been known since the very beginnings of computer use, and one of the early answers was the so called *separate compilation* – splitting of every program into relatively independent modules – so-called *compilation units* – that can be compiled separately but later on can form again one entity by means of so-called linking. Separate compilation offers a lot of advantages, *e.g.*:

- each compilation unit needs less resources for compilation than the whole system or entity;
- changes that have to be made to one module do not require compilation of the other modules;
- it is easy to organize the development of the separate units in parallel, and as a result to complete the developed product much earlier.

Is such an approach applicable to modelling in mechatronics? Yes, but there are a few prerequisites:

26. It should be possible to prepare the separated models (or, similarly to the units in the separate compilation – model units) independently from one another;

27. It should be possible to integrate the prepared model units or components after their separate authoring without remodelling or converting;

28. Assuming that a given entity has been split for separate modelling since the resources are insufficient to model (or to author) it as a whole, there should be

a guarantee that the combined (or "linked", or integrated) entity will not require more/additional resources.

So, let us see what the resources are used for and how we could utilize them better. Assume we are analysing a hypothetical platform where only the software necessary for modelling is running. Then we have an operating system and an authoring (or modelling) tool running on the respective hardware. In the case of SCA, the authoring tool is some CAx-system. In MCA it can be a CAx-system, a simple text editor or some specialized tool.

Suppose for a moment that the so called virtual memory has not yet been invented, and in order to work with a model, everything needed – BIOS, OS, authoring tool and model – has to be loaded into the main memory. Moreover, some additional space is needed for temporary/working storage. Assuming that the memory is organized linearly, we can represent this visually as follows:

BIOS	OS	Authoring tool	Workspace	Processed model

Figure 4.4. Distribution of the software in the computer memory at a given moment

Now consider the following:

- The least useful thing in memory is the workspace, but usually the workspace needed for processing a model is directly proportional to its size.
- Economists and manufacturing planners try to achieve maximal workload of all resources, which means that in the worst case the largest models either cannot be loaded at all or the workspace is insufficient and they cannot be processed.
- The largest amount of memory is usually occupied by the authoring tool (especially in the case of a CAx-system).

Now compare the typical sizes of a host and a CAx-system (graphically represented in Figure 4.11 on page 189). Assume that you have created (authored) ten equally large models – so large that even 1 byte larger would not fit in the memory for processing. Now check whether the memory would be sufficient for all ten models plus the hosts for all of them if you remove the authoring tool (it is not needed anymore). You will be surprised how much memory can be found in any old computer!

4.6.5   Organization and Architecture

A MCA-conforming model should be capable of representing not only the functionality of the modellee but also other aspects like outlook and behaviour (*cf.* requirements 0.1, 0.2, 0.3 from Table 4.2 and Table 4.4) and arbitrary attributes. This can be achieved by employing hierarchical (data) structures, organized in layers with different levels of detail (LODs). One of the prerequisites for achieving flexibility is that the building blocks of this structure ought to be components (more about flexibility in 4.6.7 below).

4.6.5.1 Components

The object models are abstract in contrast to their modellees, which are usually real objects. In the case of SCA they have a kind of formal representation (possibly even several representations). For ideal modelling (pursued by MCA) it is not enough to have simply representation of the modellee, since it would be passive. In order to satisfy requirements like R7, R11, R12 and R17, models have to be able to perform activities. This means that each MCA-model shall contain either an implementation of the related algorithms or links to their well-known or standard implementations. On the one hand, the simplest MCA-models (*i.e.* those from the lowest level of the hierarchy) have to be implemented as *components*, in the sense of the components definition of OMG and the additional three requirements, presented in Section 3.1.3.3.5 (on page 142). On the other hand, requirements such as 3.1.1 can be satisfied mainly through incorporation of intelligence into the components above a certain level in the hierarchy.

4.6.5.2 Hierarchy

The next important consideration concerns the correspondence among modellees, models and components (*cf.* also the term similarization in 4.6.7.2 below). Since numerous components already exist as implementation of different, mainly computer or computation related objects, it is rational to use them during the design, composition and implementation of our engineering models in components. Therefore, even the modelling of the simplest engineering component (*e.g.*, nut, bolt, gear, *etc.*) will be a model that is a compound software component, relying on numerous existing (software) components from different component libraries. Due to some additional requirements for these compound components, they should implement a number of interfaces (in the sense of OMG) and obey a list of conventions. We shall call the resulting pieces of software Autonomous Intelligent Entity (AIE) and classify them as the basic building block of MCA. On the next level in the hierarchy (*cf.* Figure 4.5) comes the composition of interconnected AIEs, which form an autonomous intelligent unit or module (AIM). AIM is usually a model of a whole device, aggregate or assembly group. In turn, binding several AIMs together (possibly also AIEs or other components from the lower levels) forms an intelligent system. When some of the AIEs are on other computers and are connected by simply communicating over the network, the resulting system is a distributed intelligent system (DIS). AIEs and AIMs can be nested and intermixed arbitrarily, have "loose connections" among the modules and might employ some redundancy to improve their performance, reliability or fault-tolerance.

It should be stressed that the implementation language or architecture have secondary meaning as long as models, implemented differently and separately, are able to interact/interoperate with one another. Otherwise, my practical experience confirms the ascertainments of other authors like the one of Janocha and Gandyra (1997) that the Java slogan "*Write once, use everywhere*" makes sense, especially if extended with "(reuse) forever".

4.6.5.3 Modelling Hierarchy (Holarchy)

On the one hand, according to MCA, almost every model above a certain level of the modelling hierarchy can be viewed *and* used as an *autonomous entity* (*i.e.,* a standalone whole). On the other hand, in both real life and the modelled worlds it

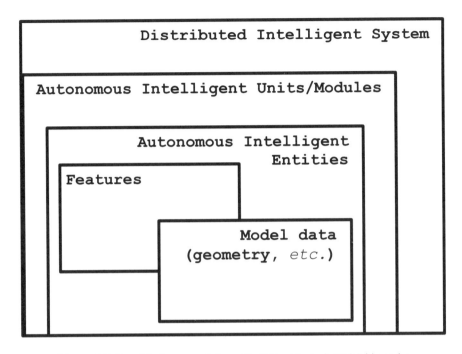

Figure 4.5. Simplified representation of building blocks in MCA hierarchy

is not clear what stands on the top of the hierarchy – the "root". We can always think of something bigger that includes or embraces everything known or defined or mentioned before. Let us give an example that "remains on the earth", although it is possible to model whole galaxies or even larger (unnamed until now) objects: assume we want to model a machine tool. It is probably placed in a workshop – the next embracing object. So we would need to model the workshop too, and we have the next embracing model. The latter is one level higher in the hierarchy but can, nevertheless, be part of yet another embracing model – *e.g.*, the model of a plant or enterprise. In this example each of the mentioned modellees or its respective model fulfils the definition of a holon (*cf.* Section 4.3.11 above).

Due to other considerations (*cf.* 4.6.7 and in particular 4.6.7.4 below) it appears quite sensible to make the autonomous entities also intelligent. Thus the autonomous intelligent entity – the main building block in MCA – has been conceived. Although it emerged as a general term that can be used on every level of the modelling hierarchy, and – technically seen – it can contain other AIEs, sometimes it is reasonable or comfortable to use names, specific to a given level – *e.g.*, product model, process model, gearbox model, machining model, *etc.* An example is presented in Figure 4.6.

Such a hierarchy – no matter whether referring to models or real objects – is a typical example of a holarchy (holonic hierarchy, *cf.* again Section 4.3.11 above). Since in Figure 4.6 the main components are, again, AIEs and since they are the main building blocks of all models in MCA, they are presented in more detail in the next section.

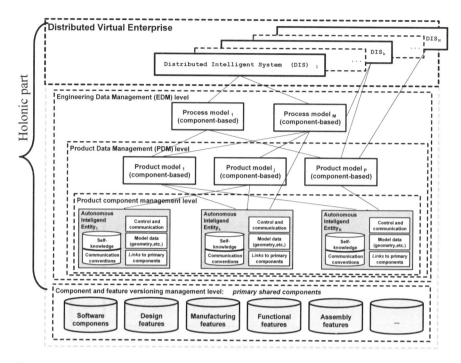

Figure 4.6. Holarchy by models of enterprises with arbitrary complexity after Avgoustinov and Bley (2003)

4.6.5.4 Autonomous Intelligent Entity

Consider the structure of an AIE of the lowest level – the product component (management) level in Figure 4.6. It is presented in Figure 4.7. Note that some modules of an AIE are case-specific – *e.g.,* model data, self-knowledge, whereas others are type-specific (*i.e.,* inherent to every AIE) – like control and communication or links to primary components (*cf.* Figure 4.6). Note also the two types of symbols representing code (rectangle) and data/information/knowledge (the disk symbol).

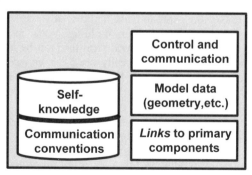

Figure 4.7. Structure on an Autonomous Intelligent Entity

4.6.5.4.1 Properties

In Avgoustinov (2004) the AIEs are briefly characterized by describing some of their inherencies as in the following (pseudo) table:

Autonomy: AIE *can take most decisions itself.*

Multi-state: AIE has the following states: *dormant* (saved on a medium, waiting); *passive* (the AIE is being edited/modified); *standby* (ready to answer requests); *learning* (acquiring additional information, which leads to new/different processing of changes in the environment) and *active* (functioning).

Persistence: AIE has a formal representation for saving on a (non-volatile) storage, obeying carefully defined but natural, preferably self-evident conventions.

Self-contained: All model-relevant data and implementation of the typical functionality are defined in the AIE itself.

(AIE)Host: Environment providing living conditions for an AIE (*i.e.* conditions for its active state).

Independence: Entering into an active state depends either only on the availability of an appropriate host or on nothing at all.

Holonomy: Typical AIEs can be used alone, used to build up other AIEs or to be built up from other AIEs (*i.e.* they are *compound*). This matches the main characteristics of a holon.

Structure: AIEs are hierarchically structured and use different levels of detail (LOD), depending on the context. Neither the level number nor their content is necessarily pre-defined.

Stratification: The aspect-related information and functionality is organized in *layers*. Viewpoints of experts in different subject areas (aspects) can be represented and used in the model.

Reflective: AIE is capable of exhibiting the most important own interfaces to (the rest of) the environment.

Procedurally representable: It is more appropriate and desirable to represent some (sets of) objects or portions of the models by parameterized procedures.

High flexibility: Achieved   through   parameterization,   expandability   and adaptability.

Parameterized: Provide a way for changing their appearance and behaviour through parameters.

Expandability: AIEs provide a way (API, reflection, documentation, *etc.*) for defining new properties, functionality, *etc.* on the base of the existing ones.

Adaptivity: An AIE can be changed without need to change the existing interfaces and can add new interfaces without need of internal changes, too.

Finally, the AIEs are harnessing the OOT and benefit from its key features like inheritance, encapsulation, data abstraction and others.

4.6.5.4.2 AIEs vs. Autonomous Agents

A frequently asked question is whether there is any difference between AIEs and agents. The short answer is yes, because an AIE models (a part of) a product or process, whereas an agent usually models or implements mainly one or more activities, which are often mapped to activities of a human.

An *agent* is defined in Howe (2006) as follows:

In the client-server model, the part of the system that performs information preparation and exchange on behalf of a client or server. *Especially in the phrase "intelligent agent" it implies some kind of automatic process, which can communicate with other agents to perform some collective task* on behalf of one or more humans.

Some important properties of an agent, according to Franklin and Graesser (1997), are given in the first four columns of Table 4.5. The last column is added in order to show which of these properties are relevant to an AIE.

Table 4.5. Autonomous agents, after Franklin and Graesser (1997) *vs.* autonomous intelligent entities (AIEs)

#	Agent Property	Other Names	Meaning	Availability of the property in AIEs
1	Reactive	(Sensing and acting)	Responds in a timely fashion to changes in the environment	Possible, but not necessary
2	Autonomous		Exercises control over its own actions	Yes
3	Goal-oriented	Proactive purposeful	Does not simply act in response to the environment	No. Object (model), purpose and aspect oriented
4	Temporally continuous		Is a continuously running process	Possible, but not necessary
5	Commu-nicative	Socially able	Communicates with other agents, perhaps including people	Communicates with other AIEs, and possibly but not necessarily with people (sensors)
6	Learning	Adaptive	Changes its behaviour based on its previous experience	Desired for the system of AIEs and for some of the AIEs, but not always necessary
7	Mobile		Able to transport itself from one machine to another	Yes
8	Flexible		Actions are not scripted	Parameters, extendibili-ty, inheritance, exchangeability
9	Character		Believable "personality" and emotional state.	Possible, but not necessary

4.6.5.5   Model Independence

One of the recommendations of MCA is to create models that have minimal dependence on other software – ideally none. Reducing the dependency on a platform (or at least on a CAx-system – *cf.* Section 4.6.3) of the models to a minimum (*cf.* requirement R8 in Section 4.2 above) can:

- reduce the accidental complexity;
- allow shifting the focus onto the most appropriate and important part of the task – the modelling (and therefore, on achieving better results);
- improve the predictability of models and systems of models;
- reduce the effort for learning/training and maintenance;
- allow the product models to be embedded in their embodiments (*e.g.*, saved on a micro-chip or RFID).

Eventually, all this leads to better efficiency.

4.6.5.6   Homogeneity and Distribution

When a system of models is built up from independent components it is often reasonable or necessary to distribute these components on different computers, perhaps even having a different geographical location. At least three important questions arise:

- If the separate components can live on different computers, could they also be integrated and how good/reliable would such an integration be?
- How do the properties of such a compound system depend on the distribution?
- How maintainable is a distributed system?

The answer to the first question is discussed in 4.6.6 below. A first approximation to an answer to the second question is presented in Figure 4.8 on the basis of well-known schemes for computer use and their traits. Note that the traits mentioned refer always to the computer on which the users are directly working, *i.e.*, on workstation, peer (*i.e.*, arbitrary network-ready computer), client and terminal. Since the boundaries between the different schemes of connection/use are fuzzy (*i.e.*, one computer can participate in multiple schemes simultaneously), the dependencies in Figure 4.8 are presented as curves rather than bar charts.

The answer to the question about the maintainability of a distributed system can be given in a similar way. Figure 4.9 presents an estimation of the maintenance effort for different schemes of computer use. Note that the domains of different possible variants within a given scheme are denoted approximately by means of the labelled grey areas beneath the x-axis.

As one can see, there seems to exist a kind of minimum in the total effort when using a client–server scheme.

4.6.6   Integration

Longstanding experiments in the fields of model exchange and integration have led the author to a somewhat surprising conclusion: no modeller has difficulties when integrating his own models, even if they are modelled in different authoring tools. It was not easy to answer why this is so, but after long investigation, the conclusion was that when a modeller integrates two or more own models:

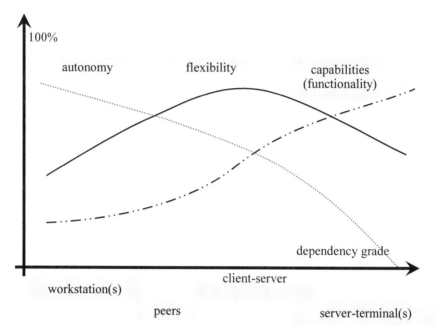

Figure 4.8. Important traits of well-known schemes of computer use depending on (de)centralization (dependency grade)

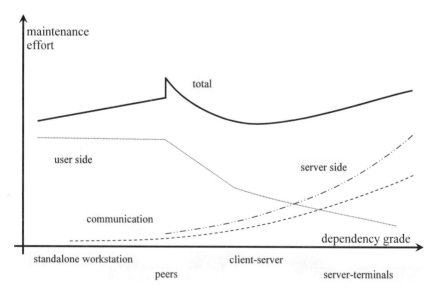

Figure 4.9. Maintenance effort for different schemes of computer use

- they are usually more homogeneous (*i.e.*, created in conformance to uniform concepts, obeing the same unwritten rules, style, *etc.*);
- he works purposefully because the integration takes place within his domain of competence;
- he knows the models and their specificities better;
- he usually proceeds with the integration bottom-up.

The latter should not be underestimated: the point is that when the smallest models are integrated first, the resulting compound models have lower interface pressure (respectively, interface resistance, *cf.* the respective definition in Section 2.4) and lower due complexity (*cf.* Section 3.1.2.4).

In contrast, most problematic are those cases where a user (*i.e.*, not the modeller himself) tries to integrate heterogeneous models by means of conversion-supported transfer of a whole model (known as model/data exchange). In this case none of the four observations mentioned above can be true: the models are heterogeneous; the work can hardly be purposeful since the integration takes place on a higher level (often beyond the boundaries of the user's domain competence); model parts are integrated regardless of their need and the foreign model is scarcely known. The integration can be described as top-down, which leads to higher uncertainty in the process flow.

Our analysis reveals the major factors: wrong or improper choice of subject, scope and standardization grade of the integration, together with insufficient qualification of the modeller/user (*cf.* Figure 3.28 in Section 3.1.3.3.2). A secondary but still important role is played by the method and the pursued extent of integration. Taken together, these factors lead to unneeded (due) complexity, lack of control, and inefficiency.

To avoid these problems, the MCA relies upon the following decisions (please, *cf.* again Figure 3.28 and Section 3.1.3.3.2):

a) choice of the functionality as integration subject (*vs.* physical integration of the components themselves or even integration of the authoring tools by SCA);

b) reduce the integration scope to a minimum – *i.e.* integrate only what is necessary (lean integration); this requires appropriate granularity of the (pursued) system of models;

c) use the minimal standardization grade (*cf.* Figure 3.5 in Section 3.1.1.12); it seems that on the lower levels, conventions not only suffice, but cause less problems than standards;

d) use event/message-driven interaction/communication as an integration method when possible; otherwise use the next preferred method – data sharing, and only if no other possibilities remain use data/model exchange;

e) never use the full extent of integration if not necessary;

f) encapsulate the resulting integrated model and define clear interfaces for its use;

g) delegate the integration to dedicated pieces of software (integrators, *cf.* Figure 4.10) and to the models themselves; the integrator is known as broker in CORBA – *cf.* Siegel (2000).

The described approach for integration achieves the shift from the upper system levels down to the model, sub-model and component levels and offers more advantages in almost all aspects.

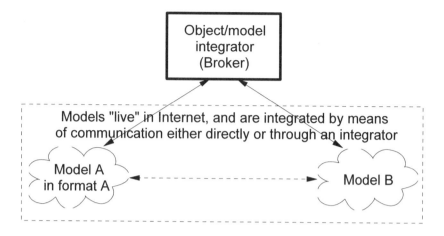

Figure 4.10. Component-based model integration after Avgoustinov)

4.6.7   Achieving High Flexibility

Model flexibility is a trait that is much sought-after but not always available to the desired extent. The more sophisticated or specialized a model is, the lower is its flexibility. Considering that most models are compound or, actually, even systems of models, it is important to know what the flexibility of such models depends on and how it can be improved.

The analysis of different existing CAx-systems, computer systems and systems of models has led us to the definition of four main (groups of) factors that have immediate impact on (model) flexibility. These are model organization, extensibility, modelling for reuse and knowledge-based, distributed decision taking. The MCA relies upon optimal use of all these factors and especially on the intensive use of the last one.

4.6.7.1 *Model Organization*
The flexibility of a model requires no radical change of everything in it. On the lowest level (*cf.* Figure 3.3 in Section 3.1.1.9) it is possible to change the values of simple variables or a data structure. On the second level it is possible to exchange a component (*i.e.*, part of the model content). On the third level it is possible to change the structure of a model or the connection/relations among components. The fourth level allows change in the functionality. The last two levels are related to changes in the organization of a compound model, changes in the paradigm or in the application concept. Closed systems cannot be changed, therefore, they are not as flexible as open systems. Both organization and granularity influence the flexibility of a sophisticated model. Conventional CAx-systems do not achieve optimal model organization, optimal model granularity, and optimal model integration. If we assume that, in general, the functionality of a software component is proportional to its "size", Figure 4.11 may give us an impression about the functionality of different software components involved in the engineering processes.

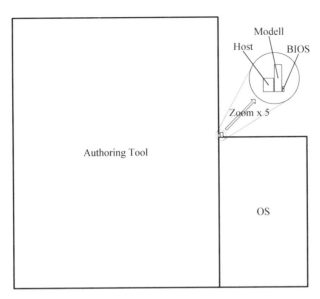

Figure 4.11. The size of an average model compared to the size of other software. The area of each rectangle is proportional to the average size of the respective software after Avgoustinov *et al.*). Note that an area of the picture is zoomed.

Figure 4.11 illustrates the huge disproportions of the mentioned software subsystems in size and, respectively, in functionality, but it still does not reveal that there seems to exist a kind of logarithmic relation between them. Another representation of the same data, illustrating this aspect, is given in Figure 4.12.

Obviously, although the CAx-systems could benefit very much from separate modelling, they cannot apply it.

4.6.7.2 Modelling for Reuse

The possibility for reuse is one of the most important traits of any piece of software. Very often the advantages that can come from reuse are either underestimated or simply not known. Therefore, we shall discuss at least some of them. Consider the ratio between the time needed to create a given artefact and the time this artefact can be used: the lower the ratio is, the higher is the economical efficiency. Ideally, the creation of an artefact should be instantaneous and its use should be time unrestricted (*i.e.,* infinite). In a real case the fastest method to create an artefact is to replicate it. The replication of material objects requires some effort, but this effort is much smaller than the effort for design, development and production taken together. The replication of software is almost as quick and effortless as in the ideal case, but the development of software is a long and expensive process. For this reason, it is very important to ensure high grade of reuse. And in the case of software artefacts (respectively, software models) two additional, software-specific methods for ensuring/supporting reusability exist: factoring out and parameterization.

To factor out a piece of data or program code means to recognize a piece that tends to recur, to separate it as a module, to test it and to make it accessible to other pieces of code. The reusability of a piece of code becomes higher when it is

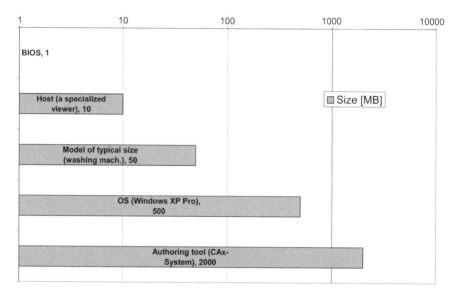

Figure 4.12. The size of an average model compared to the sizes of other software (logarithmic scale)

possible to make it usable not just for a specific task, but for a whole class of similar tasks, instead. This is achieved through parameterization.

Some of the prerequisites for reusability were mentioned in 2.4.1.34 and Figure 2.31. In the case of software models, there exist some additional factors (or modelling parameters) that can increase the reusability – similarization, intuitiveness, mnemonics, self-documentation/self-awareness.

The similarization means that a newly created model resembles not only the outlook of the modellee, but also its behaviour and the way it interacts with the user and with the environment. For instance, if a model of a light switch is modelled, the change of its state (on/off) should be controlled not by means of a menu (unintuitive) but by touching it with either the mouse-pointer or with the hand of the user's avatar[43].

Achieving intuitiveness is extremely important for modelling, but very different in SCA and MCA. Working with a (unknown) CAx-system, one contemplates "how could I achieve this?", and tries to guess in which menu the respective command is. Working with an unknown MCA-model of a known product or object the respective question is "how would I do the same with this model's modellee?". The authoring of intuitive models requires a respective convention and a more concentration on the modelling process. The easiest way to achieve this is to shift the focus from the authoring tool on the modelling itself. To make (a part of) a model intuitive means that its potential user would know how to deal with it even before/without reading the documentation. This is often easily achievable, but underestimated and neglected. The use of mnemonic techniques in the modelling

---

[43] Avatar is used here with the meaning of (virtual) model of the user, representing him in a modelled world.

makes it easier to remember different attributes, traits or activities of a model or how to interact with it.

Finally, the self-awareness and self-documentation facilitate the dealing with unknown models. Their importance is proportional to the functionality of the respective model and increase with its lifetime. And since the software does not age physically, we are pursuing long living models. Therefore, the best way to ensure their usability throughout their whole life is to incorporate their documentation into them.

4.6.7.3 Extensibility

In its easiest form, the extensibility can be viewed as a convention to develop the models in such a way that they can be extended in order to increase their adaptability on demand. This means that if (the sources of) the models are closed (unreadable, unchangeable, etc.) they should expose some mechanism for extension like interface description, documentation, etc. The MCA requires in explicit form clearly defined interfaces of all models and components and recommends embedding some form of documentation into the models and components themselves.

4.6.7.4 Knowledge-based Distributed Decision Taking

The incorporation of intelligence into software components and models offers huge potential for increasing the flexibility by enabling the autonomy of models on the middle levels in the model hierarchy, and integration of the functionality of models instead of integrating the models themselves. Avgoustinov and Bley (2006) argue that the main way to achieve intelligence is the incorporation of (event-driven) behaviour and (self-)knowledge. They affirm that intelligence can be built only upon (domain) knowledge, and adduce that this knowledge has to be incorporated into the models in a bottom-up manner for the following reasons:

- knowledge builds upon information, and information is data in a context;
  - if we begin top-down, there exists only one term (single data?) and no context at the beginning;
  - therefore, the bottom-up method seems more natural and advantageous;
- knowledge comes from humans' brains, but they express it word-by-word, from pieces to the whole, also meaning bottom-up;
- knowledge acquisition and representation is a time and labour consuming process ➔ reuse of even intermediate results makes sense;
- it is easier to reuse smaller parts of a knowledge puzzle than to reuse the whole, especially if the whole is still not ready.

Thus, on the basis of some basic knowledge an intelligent model:

- allows the decision taking to move to a lower modelling (or engineering) level, where the complexity is lower;
- allows decisions to be taken near to the cause/need for the decision or to pass it upward in the hierarchy when the locally available information is insufficient for a decision;
- is self-aware and can answer questions concerning it (e.g., "what is your name", "what are you capable of", "are you free", "do you have free resources", "can you perform X", etc.;
- has the ratio of needed instructions to accomplished work tending to zero;

- can complete some simple self-maintenance activities;
- is capable of learning;
- can represent both static and dynamic data (*i.e.* product and process related).

Therefore, MCA not only relies on use of knowledge and intelligence, but recommends incorporating them directly where they are needed. A simplified version of the algorithm used to create intelligent models is presented in Figure 4.13.

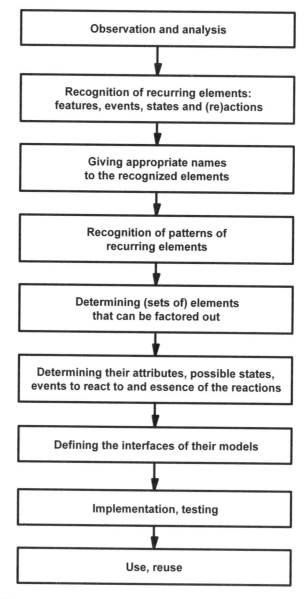

Figure 4.13. The steps towards incorporation of intelligence into models, after Avgoustinov and Bley (2006)

It is important to note that when following this approach, sooner or later – depending on the model granularity (*i.e.*, on the models' average size) – name collisions can occur. These are normal for bottom-up development and formally could be avoided with careful choice of all names and suitable application of namespaces. The real problem, though, is not how to avoid a name collision in the technical sense, but how to ensure that in a (future) world with unlimited number of models, exactly the desired model is unambiguously referred to. On the other hand, the opposite problem is also possible: to know that a model with the desired qualities exists, but to be unable to find it because its name is unknown. These two problems will receive a satisfactory solution only with the development of taxonomies and domain ontologies for the concerned domains. The introduction of domain ontologies would also improve the development of the user interfaces, as well as the human-to-human communication.

With respect to globalization and the international cooperation we consider also very important the employment of thesauri. It would increase the comfort of use and the probabilities for finding the searched. According to our approach in Avgoustinov and Bley (2003) the intelligence is distributed within the process model (in our case, of assembly and assembly planning) across several hierarchy levels. The behaviour making the models appear intelligent is implemented on the lowest level with the help of features, patterns and procedures, integrated into autonomous intelligent entities (AIEs). On the upper levels the AIEs can be combined in more and more sophisticated models, up to distributed intelligent virtual enterprises. Very interesting possibilities are offered by the combined use of (assembly) patterns and (assembly) features – they have many similarities, but also many differences that can nicely complement each other.

Summing up, with the concepts for separation of authoring from use, separate modelling, model organization according to Figure 4.6 and incorporation of intelligence, the MCA offers much higher flexibility of the prepared models.

4.6.8   Cooperative Work and Distributed Authoring

The separation of authoring from use, together with the separate modelling, create prerequisites for easier cooperation, since it makes no difference where the models are created. When the cooperating authors (or modellers) are distributed in different points of the world we can speak of *distributed authoring*. An illustration of how cooperation on the basis of distributed authoring can be organized is presented in Figure 4.14. Note the difference between Internet and (manufacturer's) intranet. The idea is that instead of using the conventional electronic catalogues it would be much more convenient if the manufacturers of different equipment – in this example workpieces, tools, machine tools, and fixtures – provide functional models of their products to be used for testing and planning of the future production. Of course, these models can involve some kind of protection against theft of intelectual properties – e.g., authorisation of the use only after identification, or even for a small fee. Nevertheless, the possibility to experiment and play different scenarios with the whole equipment and prove that it is exactly what is needed can enormously improve the productivity, the economical efficiency and probably even the quality of the production yet during its planning. And such way of cooperation could provide for much earlier and more intensive feedback concerning the quality of the involved models and their respective modellees.

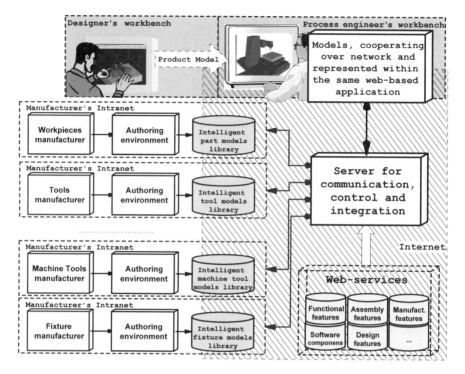

Figure 4.14. A scheme for cooperative work over Internet, advanced from Avgoustinov (1999) and Avgoustinov and Bley (2003)

4.6.9   Lean Modelling

As already mentioned in Section 4.6.1, MCA can be viewed as a new organisation of the modelling and the resulting models. This new organisation could play for the modelling the same role, which the concept of lean manufacturing, Womack and Jones (1994), lean thinking, Womack and Jones (2003) and lean solutions, Womack and Jones (2005) played during the last decades for the manufacturing: minimising the waste and increasing the efficiency.

And if the approach is successful, it is not important, whether its name is lean modelling, model centred approach or something else; important is to make the next step on the way of continuous improvement.

4.7   Comparison of MCA and SCA

A comparative assessment of the most important traits of the product and process
models, developed by means of the two approaches, is presented in Table 4.6 and
Table 4.8. The comparison is intended just to give some orientation and first
impressions about the MCA.

For SCA, the theoretical possibility for a certain parameter to be employed
(either 0 or 100%) and an estimation of how many CAx-systems are making use of
it really are given. The value 5% denotes that probably there is a CAx-system,
which makes use of this parameter, but it is not known to us.

Table 4.6. Modelling parameters (strategic and conceptual) in use by SCA and MCA (*cf.*
Table 4.4, UL part). Parameters with the same use in both SCA and MCA are not shown.

			SCA		MCA	Correlation
			theoret. possible	in practice	theoret. possible	
Modelling parameters	strategic & conceptual activities	separation of authoring from use	100%	5%	100%	1
		shifting the focus from use of tools towards adequate models/results!	0%	0%	100%	2
		(modellee) interaction resemblance	100%	30%	100%	3
		parameterization	100%	60%	100%	4
		pursuing optimal granularity		5%	100%	5
		platform-independent model design	100%	5%	100%	6
		integra-bility interoperability	100%	10%	100%	7
		standardization	100%	80%	100%	8
		approp-riate hosting long living hosts	100%	50%	100%	9
		high performance	100%	80%	100%	10
		low price of the host	0%		100%	11
		new versions backward compatible	100%	75%	100%	12
		use of patterns	100%	10%	100%	13
	choice of appropriate authoring tools	tool performance	100%	70%	100%	14
		quality of the delivered models	100%	50%	100%	15
		tool lifetime	100%	70%	100%	16
		tool reliability	100%	90%	100%	17
		new versions backward compatible	100%	75%	100%	18
		low price of the tool	100%	5%	100%	19

The column "in practice" is not given for MCA, because it is still in its
introductory phase. Neither Table 4.6, nor Table 4.8, which is the continuation, can
give a final, qualified answer to the question of which approach is better. What

they are giving you is a basis for comparison, showing which modelling parameters lead to differences when used/changed.

Table 4.7. Modelling parameters (concerning organization and implementation) in use by SCA and MCA (*cf.* Table 4.4c and Table 4.4e). Parameters with the same use are not shown.

| | | | | | SCA | | MCA | Correlation |
					theoret. possible	in practice	theoret. possible	
Modelling parameters	model organisation	architecture	compo-nent-based	clear interfaces	100%	5%	100%	20
				libraries	100%	5%	100%	21
				broker	100%	5%	100%	22
				open	100%	5%	100%	23
				modular	100%	5%	100%	24
				fractal	100%	5%	100%	25
				decentralised	100%	5%	100%	26
		autonomy and intelligence	incorpo-ration of self-awareness	capability to answer predefined questions about itself	0%		100%	27
				capability for self-maintenance	0%		100%	28
			incorpo-ration of intelli-gence	capability to take decisions	0%		100%	29
				capability for self-improvement	0%		100%	30
			max. independence from the host		0%		100%	31
		structure		hierarchical	100%	80%	100%	32
				layered	100%	70%	100%	33
Modelling parameters	implementation modalities	object-oriented	allow user-defined types		100%	30%	100%	38
		employment of recursion			100%	5%	100%	39
		proper choice of the level of standardization			0%		100%	40
		minimization of the information content			0%		100%	41
		mnemonics			100%	60%	100%	42
		factoring out reusable content			100%	40%	100%	43
		user qualification			high		low	45

These two tables could be used together with Table 4.4 for determining which customer demands would be influenced. An attempt to evaluate which of the two approaches is better suited to a particular situation can be carried out after introducing weight factors for the modelling factors or even better – for the (adapted to the specific situation and needs) functional requirements. We leave this task as an exercise to you – probably after considering the remainder of the book.

4.7.1 Modelling Efficiency

Since most systems have a hierarchical structure, the lower in the hierarchy we go the higher is the number of the components on the respective level and the lower is their complexity – *cf.* Equation 3.10 and Figure 3.16. Let us consider how this can influence the modelling efficiency.

The efficiency of a process is usually defined as the ratio of its output to its input. The output of the modelling process is a model. Assuming that the requirements for the model define the modelling process, but do not belong to it, what remains on the input are the modelling techniques and tools. Therefore, the modelling efficiency for the creation of just one model can be expressed as follows:

$$efficiency_{modelling} = \frac{value_{modell}}{price_{techniques} + price_{tools}} \qquad (4.1)$$

According to this equation (which neglects the value of time savings) the efficiency would tend to infinity when the divisor tends to zero. Supposing that smaller and less complex models have lower value, it appears more efficient to use cheap tools and techniques for small models, and more expensive tools and techniques for larger, more complex models. The only fact that should not be disregarded is that the use of more than one tool for authoring can cause incompatibilities or problems during the integration of the separate models.

Of course, no tool is used for only one model. To take this into consideration we should estimate how many models (denoted by N below) could be produced during the lifetime of the respective tool and what their value is.

$$efficiency_{modelling} = \frac{\sum_{i=1}^{N} value_{modell_i}}{price_{techniques} + price_{tools}} \qquad (4.2)$$

If we assume that the frequency of the need to edit (*i.e.*, create or modify) a component (without its sub-components!) does not depend on its size, and that each modification is performed within a CAx-system, it follows that each system will be used more frequently to modify small components than large ones.

4.7.2 Systems for Modelling (Authoring Tools) *vs.* (Systems of) Models

In Table 4.8 is presented a short comparison between the systems for modelling and groups of models, integrated into systems.

Table 4.8. Systems for modelling *vs.* systems of models: a short comparison

Trait	Systems for modelling	Systems of models
Aim	To satisfy the needs within a given domain (abstract aim!); needs are permanently growing. Moreover, to be flexible the systems have to posses reserve functionality➔ another cause for growing.	Each model should be, by definition, adequate but still finite representation of the modellee (concrete/tangible aim). A system of models is a (finite) number of models.
Aim abstractness	abstract/fuzzy➔ more difficult to pursue.	concrete/tangible➔ much easier to pursue.
New version	shows how far from perfect the system is.	brings the model towards its aim by making model-based decisions as sensible as the modellee-based.
Growing	every system tends to get more and more functionality and nothing stops it from growing infinitely.	Models must remain finite (according to their definition!); the size of compound models (systems of models) can be huge, but it still remains a sum of finite number of components, each of finite size.

From Table 4.8 it follows that when (CAx)systems are used as both tools for modelling *and* hosts for the created models, they have a much smaller chance of becoming perfect than systems of models hosted independently on lean hosts. Consequently, it is better to focus on the development of the models themselves, which is one of the main ideas of MCA.

5

Conclusion

A new scientific truth does not triumph by convincing opponents and making them see the light, but rather because its opponents eventually die, and a new generation grows up that is familiar with it.

Max Planck
A scientific Autobiography and Other Papers, 1949

Whenever science makes a discovery, the devil grabs it while the angels are debating the best way to use it.

Alan Valentine

The way to achieving The Perfect Modelling approach can be very long, even endless. But as the saying puts it, even the longest way starts with the first step. The Model Centred Approach described in the previous chapter is an attempt not only to make the first step on this way, but also to achieve some advancement towards a better – and simultaneously towards The Perfect – modelling approach.

5.1 Based on Highly and Easily Integrable AIEs

One of the main problems of SCA – the integration of model components created in different CAx-systems – is solved in MCA by unique shifting of the integration paradigm. In contrast to SCA, which relies on integration of the component themselves, or even worse – on integration of the authoring tools to achieve simply integration of the model components – MCA relies on *integration of the component's functionality*, accomplished by means of cooperative work and communication (for details on different types of integration we refer to Section

3.1.3 above). This concept for integration resembles very much how humans cooperate to integrate their efforts and is based on three pillars: design for integration, componentization and proper choice of the level of standardization.

Alternatively, backed by Knowledge-based Adaptive Conversion proposed in Avgoustinov (1997) it can support hierarchically-distributed (*i.e.*, separately implemented for each model or component and not as a standalone converter) offline and online exchange and thus allow the adoption and reuse of models, created by means of distinct CAx-systems.

5.2   Extremely Flexible and Extendable

Due to its open organization and architecture, the flexibility of MCA remains high even when the model components are not open themselves. The organization, having no fixed structure and based on components with clearly defined interfaces, enables easy exchange of separate components, changes in the interaction of the model components, exclusion of unneeded or inclusion of additional models. This allows even benchmarking of components with the same purpose and interfaces, but from different producers, and to choose/buy the one, which is best suited to the respective purpose.

The proposed incorporation of intelligence in components at the middle and upper levels of the model hierarchy (AIEs, AIMs, *etc.*; Figure 4.6) increases the flexibility and efficiency even more, unburdening in addition both model authors and model users from the frequent need to make routine decisions, calculations or other recurring routine activities.

The high flexibility and extendibility make MCA universal and useable in a number of application areas. Combined with its efficiency, this makes it well-suited also to small and medium-sized enterprises.

5.3   Network and Web-ready Capabilities

Another advantage of MCA is its readiness for use over networks, and possibilities to support distributed cooperative work. There are two aspects here. On the one hand, humans are supported in their cooperation according to at least the following combinations:

- author to author: for cooperative development;
- author to user: for teaching and training;
- user to user: for exchanging experience or cooperative work/use; and
- user to author: for feedback, request of new models/functions or other customer support.

On the other hand, it is possible to compose sophisticated models, whose components are distributed in diverse geographical locations but can be used over the network just as easily and conveniently as the local use of any software system would be.

5.4 Scalability up to and Beyond Distributed Virtual Enterprises

The open organization and the easy integration and flexibility offered by MCA, satisfy the prerequisites for achieving arbitrary scalability of the models. Their network-readiness, in addition, provides a way to scale their performance by distribution of the model parts and the diverse calculations on different computers and other hardware resources.

5.5 Advantages for Modelling

Summing up, compared to SCA the MCA offers plenty of advantages – among them higher intuitiveness and ease of use, low (accidental) complexity, leading to increased efficiency of the modelling, high flexibility, interchangeability and extendibility of the sub-models, low costs for development, use and maintenance and reduced volumes of information flow. All these taken together lead to higher reusability, and eventually to higher economic efficiency.

6

Perspectives

Science is always wrong. It never solves a problem without creating ten more.

George Bernard Shaw

Exactly as George Bernard Shaw puts it, the MCA does not only offer a solution, it raises also questions and creates some problems. One of the weak points is that the increasing number of components in a model leads to more effort on management and organization. Typically this can be solved through proper encapsulation, but maybe not always. On the other hand, the product data management (PDM) systems and the product lifecycle management (PLM) systems can do a great job here, but what they cannot do is the integration of the separate components – unless these model components possess enough intelligence to either integrate themselves alone, or support the PDM/PLM system in this activity.

For the evaluation of MCA in more detail, a library of prototype AIEs, AIMs and parameterized models of numerous representative standard parts and manufacturing tools was prepared. This library was successfully used for the modelling of different products and processes in engineering areas like assembly, presented in Bley *et al.* (2002), Avgoustinov *et al.* (2006), Avgoustinov and Bley (2006) and Bley *et al.* (2006); machining simulation, presented in (Avgoustinov (2000a), Avgoustinov (1999), Avgoustinov (2000a); Avgoustinov and Bley (2000) and Avgoustinov (2000b); education, presented in Avgoustinov (2000b); layout planning and others, presented in Avgoustinov and Bley (2003), Avgoustinov and Bley (2004) and Avgoustinov and Bley (2005). The overall impression from the testing conducted on this early stage of development is that the MCA performs remarkably well and according to expectations. The proposed method for model integration reduces the interface pressure and establishes good prerequisites not only for data exchange and data sharing, but also for online model integration in real time. The flexibility, extendibility and ease of use of the AIEs and AIMs – even over Internet very attractive are. But one of the most important MCA features is without doubt the *change in the way of thinking*, revealing a completely novel perspectives and leading also to a change in the modelling workflow as symbolically represented in Figure 6.1.

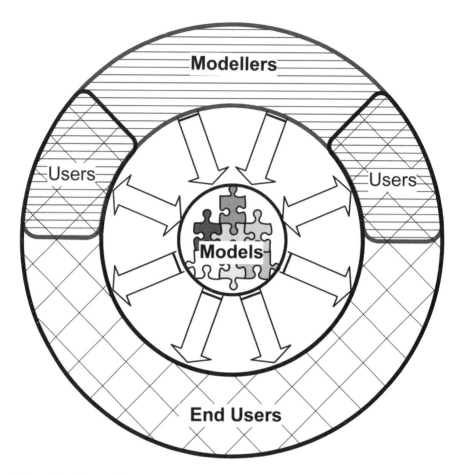

Figure 6.1. The Modelling Wheel: a schematic representation of the workflow in the Model Centred Approach

The price of any MCA-component is expected to be incomparably lower than that of any CAx-system. It is highly probable that even the total price of the most complicated model, prepared using MCA and including the prices of all necessary AIEs, AIMs, standardized parts models, hosts and other components, will be lower than the price of an average CAx-system. At first glance, this seems not very attractive to CAx-systems producers, because it seems easier to make revenue with big, expensive systems and the support for them, than with small components and their support. But this situation resembles the situation in the computer market at the time of the appearance of the first personal computers: initially many producers of mainframe computers could not see a benefit in selling cheap PCs. Later on, it turned out that these much smaller and cheaper computers are affordable for much higher numbers of customers – including private persons – and they conquered the market. Since the lower price has led to a high number of PCs sold and their lifespan is shorter than that of the mainframes, the number of PCs produced continues to increase. Without this still continuing increase we would not have the achievements made by computer science during recent decades, which many experts call "revolutionary". And nowadays the revenues from sales of personal computers, software for them and consulting related to them is considerable. Analogously, the intensive use of autonomous and intelligent models could be revolutionary for modelling in a similar way and can be useful both for its users and for the carriers of know-how about it.

We expect that the core features of MCA like flexibility, portability, connectivity, autonomy, web-readiness, easy maintenance and low price will make it popular among end users if a software company takes over the technology and its support and further development. Similarly to PCs, which are capable of accomplishing many tasks alone but for other tasks just complement the mainframes and supercomputers, the MCA-components will be able to accomplish numerous tasks alone and for others, will complement different CAx-systems. Although the final proportion between MCA and SCA is still an open question, we have no doubt that – in one or other form of implementation – the MCA will establish itself as a reasonable and efficient complement and enhancement of the SCA.

7

Afterword

> *The scientist is not a person who gives the right answers, he's the one who asks the right questions.*
>
> Claude Lévi-Strauss
> Le Cru et le cuit, 1964

Probably not every explanation in this book was absolutely clear. It is even possible that not everything is absolutely correct – at least from the viewpoint of your (the reader's) specific background and qualification. Nevertheless, I would like to hope that you have liked at least some ideas, or that you at least do not disagree with everything.

Anyway, since you are reading this, you should have at least some experience with some CAx-system, I guess. Therefore, independently from the fact whether I did gave any answers to any of the questions that have bothered you, let me finish in the sense of Claude Lévi-Strauss (*cf.* the motto) and finish this book with a few questions:

1. Do you still believe that the time of computer integrated systems (CIM) in the sense of integrated systems for authoring, will ever come?

2. Do you still prefer to use models within their huge, expensive and complex authoring tools, or would a simple host suffice?

3. Do you prefer to focus your effort and resources on the (authoring) tools instead of on the problems you have to solve?

4. Do you think that the System Centred Approach in its current state will still be The Approach of the twenty-x[th] century?

If your answer to at least one of these questions is yes, I would be glad and grateful to hear from you.

Glossary and Used Abbreviations

AIE	Autonomous Intelligent Entity
AIM	Autonomous Intelligent Model
AIS	Autonomous Intelligent System
API	Application Programming Interface
BIOS	Basic Input-Output System
CAD	Computer Aided Design
CAE	Computer Aided Engineering
CAM	Computer Aided Manufacturing
CAP	Computer Aided Planning
CAPP	Computer Automated Process Planning
CATIA	Computer Aided Three-dimensional Interactive Application
CAx	Computer Aided …
CIM	Computer Integrated Manufacturing
CMM	Coordinate Measuring Machine
CNC	Computer Numerical Control
CORBA	Common Object Request Broker Architecture
CWM	Common Warehouse Metamodel
DCOM	Distributed Component Object Model
DIS	Distributed Intelligent System
DMU	Digital Mock-up
EAI	Enterprise Application Integration
ENX	European Network Exchange
HLA	High Level Architecture
EDM	Engineering Data Management
FEM	Finite Element Method
FMS	Flexible Manufacturing System
GIS	Giant Scale Integration
HMS	Holonic Manufacturing System
IC	Integrated Circuit
iViP	Integrated Virtual Product Creation
LOD	Level of Detail
MCA	Model Centred Approach
MDA	Model Driven Architecture

MEM	Mechanical Engineering and Mechatronics
MIS	Management Information System
MOF	Meta Object Facility
MOM	Message-Oriented Middleware
NC	Numerical Control NC
OEM	Original Equipment Manufacturer
OOP	Object-Oriented Programming; Object-Oriented Paradigm
OOT	Object-Oriented Technology
OS	Operating System
OSI	Open Systems Interconnection
OMG	Object Management Group
ORB	Object Request Broker
PDM	Product Data Management
PDML	Product Data Markup Language
PDGL	Part Design Graph Language
PLM	Product Lifecycle Management
RFID	**R**adio-**f**requency **id**entification; **R**adio **F**requency **I**dentifying **D**evice
SCA	System Centred Approach
SLS	Selective Laser Sintering
SME	Small And Medium-Sized Enterprise
STEP	**St**andard for **E**xchange of **P**roduct **D**ata
UML	Unified Modelling Language
VAT	Value-Added Tax
VLSI	Very Large-Scale Integration
XMI	XML Metadata Interchange
XML	**Ex**tensible **M**arkup **L**anguage

Bibliography

"ANIKA" (1998): Further information about ANICA, ProDMU, the CAx object bus and CAx components. http://rkk.mv.uni-kl.de/ComponentCAx/ComponentCAx_engl.html, Last visited: 2002

Abramovici, M., Gerhard, D. and Langenberg, L. (1998): Supporting Distributed Product Development Processes with PDM. In: Proceedings of New Tools and Workflows for Product Development - CIRP Seminar STC Design, May 1998, pp. 1–11, Fraunhofer IRB, Berlin

Arnold, F., Janocha, A. T. and Swienczek, B. (1999): Rendezvous der Monolithen: Integration heterogener CAx-Systeme. *OBJEKTspektrum,* Vol. 6, pp. 22–27

Avgoustinov, N. (1997): *Minimizing the Labour for Exchange of Product Definition Data Among N CAx-Systems.* Saarland University, Saarbrücken

Avgoustinov, N. (1999): Virtual Shaping and Virtual Verification of NC-Programs. *CIRP - Journal of Manufacturing Systems,* Vol. 29, No. 4/99, pp. 287–292

Avgoustinov, N. (2000a): Implementation of four-dimensional machining simulation in virtual reality over Internet. In: McGeough, J. A. (Ed. *Computer-Aided Production Engineering,* pp. 105–113, Professional Engineering Publishing, London

Avgoustinov, N. (2000b): VRML as means of expressive 4D illustration in CAM education. *Future Generation Computer Systems*, 17, pp. 39–48

Avgoustinov, N. (2002): Product and Process Modelling: System-centred vs. Model-centred Approach. In: Proceedings of 35th CIRP International Seminar on Manufacturing Systems, 13–15 May, Seoul

Avgoustinov, N. (2004): Product and Process Modelling: System-centred vs. Model-centred Approach. *CIRP - Journal of Manufacturing Systems,* Vol. 33, 5, pp. 437–444

Avgoustinov, N. and Bley, H. (2000): Method for reduction of machining time by the simulation of NC-Programs in virtual reality. In: McGeough, J. A. (Ed. *Computer-Aided Production Engineering,* pp. 123–129, Professional Engineering Publishing, London

Avgoustinov, N. and Bley, H. (2003): Distributed Virtual Enterprises (DVE) in Modelling, Simulation and Planning. In: Proceedings of Progress in Virtual

Manufacturing Systems, 3–5 June, pp. 123–129, Saarland University, Saarbrücken

Avgoustinov, N. and Bley, H. (2004): Web-based Process Modelling for the Digital Factory. *CIRP - Journal of Manufacturing Systems,* Vol. 33, 1, pp. 77–82

Avgoustinov, N. and Bley, H. (2005): Towards Integration by Design. In: Proceedings of International PACE Forum Digital Manufacturing, Partners for the Advancement of Collaborative Engineering Education (PACE), Darmstadt

Avgoustinov, N. and Bley, H. (2006): Incorporation of Computational Intelligence into Assembly Planning Software. In: Proceedings of Intelligent Computation in Manufacturing Engineering, 25–28 Juli, pp. 497–502, University of Naples Federico II, Ischia (Naples)

Avgoustinov, N., Bley, H. and Weyand, L. (2007): Influence of the Software Models' Intelligence on their Flexibility, Reconfigurability and Agility. In: Proceedings of International Conference on Changeable, Agile, Reconfigurable and Virtual Production, 22–24 July, Toronto

Avgoustinov, N., Bossmann, M. and Bley, H. (2006): Supporting the Assembly Planning by Means of Features and their Influence on the Development Process. *CIRP - Journal of Manufacturing Systems,* Vol. 35, No 3, pp. 257–266

BEA Systems, I. (2004): BEA WebLogic Rapid Business Integration. http://www.bea.com/content/news_events/white_papers/BEA_WL_Integration _ds.pdf, Last visited: 2006

Black, P. E. (2005): Dictionary of Algorithms and Data Structures at the National Institute of Standards and Technology (NIST). http://www.nist.gov/dads/, Last visited: 2005, National Institute of Standards and Technology (NIST)

Bley, H., Avgoustinov, N. and Franke, C. (2002): Towards Feature-based Intelligent Assembly. *Annals of the German Academic Society for Production Engineering,* Vol. IX, 1, pp. 97–100

Bley, H., Avgoustinov, N. and Zenner, C. (2006): Assembly Operation Planning by Using Assembly Features. In: Westkämper, E. (Ed. *First CIRP International Seminar on Assembly Systems,* pp. 115–121, University of Stuttgart, Stuttgart

Bley, H. and Franke, C. (2001): Integrating Product Model and Production Model in the Digital Factory. *wt Werkstattstechnik,* Vol. 91 h4, pp. 214–220

Bochmann, O., Brussel, H. V. and Valckenaers, P. (2003): *Micro and MacroLevel Interactions in MultiAgent Manufacturing Systems.* Saarland University, Saarbrücken

Bollinger, J. G. (Ed.) (1998): *Visionary Manufacturing Challenges for 2020.* National Academy Press, Washington

Booch, G., Rumbaugh, J. and Jacobson, I. (1999): *The Unified Modeling Language User Guide.* 6 edn, Addison Wesley Longman Inc., Reading

Brooks Jr., F. P. (1987): No Silver Bullet: Essence and Accidents of Software Engineering. *Computer,* April

Bullinger, H.-J. (Ed.) (2007): *Technologieführer.* Springer, Berlin Heidelberg

Burkett, W. C. (1998): PDML: Product Data Markup Language - A New Paradigm for Product Data Exchange and Integration. http://www.pdml.org/whitepap.pdf, Last visited: 2002

Ciupke, O. and Schmidt, R. (1996): Components as Context-Independent Units of Software. In: Proceedings of Special Issues in Object-Oriented Programming. Workshop Reader of the 10th European Conference on Object-Oriented Programming ECOOP, Juli 1996, Linz

Dankwort, C. W. (1997): CAx System Architecture of the Future. In: Roller, D. and Brunet, P. (Eds) *CAD Systems Development, Tools and Methods*, pp. 20–31

Dörner, D. (1987): *Problemlösen als Informationsverarbeitung.* 3. edn, Kohlhammer, Stuttgart

Duffy, A. H. B. and Andreasen, M. M. (1995): Enhancing the Evolution of Design Science. In: Hubka, V. (Ed. *International Conference on Engineering Design (ICED)*, pp. 29–35, Heurista, Praga

Edmonds, B. (1999a): Syntactic Measures of Complexity (Thesis). University of Manchester, Manchester

Edmonds, B. (1999b): What is Complexity? - The philosophy of complexity *per se* with application to some examples in evolution. In: Heylighen, F. and Aerts, D. (Eds) *The Evolution of Complexity*, Kluwer, Dordrecht

European Union, C. o. t. (2003): COMMISSION RECOMMENDATION of 6 May 2003 concerning the definition of micro, small and medium-sized enterprises. *Official Journal of the European Union,* Vol. 46, L 124/36, pp. 36–41

Fagade, A., Kazmer, D. and Kapoor, D. (1998): A Discussion of Design and Manufacturing   Complexity. http://http://kazmer.uml.edu/Staff/Archive/XXXX_Design_Manufacturing_Co mplexity.pdf, Last visited: 2005

Fischer, K., Schillo, M. and Siekmann, J. (2003): Holonic Multiagent Systems: A Foundation for the Organisation of Multiagent Systems. In: Marík, V., McFarlane, D. C. and Valckenaers, P. (Eds) *First International Conference on Applications of Holonic and Multiagent Systems (HoloMAS'03),* pp. 81–90, Springer, Prague

Frankel, D. S. (2001): Model-Driven Architecture™ Reality and Implementation. http://www.IONA.com, Last visited: 2004, IONA Technologies, Inc.

Franklin, S. and Graesser, A. (1997): Is it an Agent, or just a Program?: A Taxonomy for Autonomous Agents. In: Proceedings of Third International Workshop on Agent Theories, Architectures, and Languages, pp. 21–35, Springer

Frizelle, G. and Suhov, Y. M. (2000): An entropic measurement of queueing behaviour in a class of manufacturing operations. In: Proceedings of Proceedings of the Royal Society, pp. 1579–1601, The Royal Society, London

Gausemeier, J., Hahn, A. and Schneider, W. (1996): Kooperatives Modellieren auf Basis transienter Objekte. In: Proceedings of Fachtagung der Gesellschaft für Informatik, 7–8 March 1996, pp. 311–325, Ruland, Bonn

Gausemeier, J., Lindemann, U., Reinhart, G. and Wiendahl, H.-P. (2000): *Kooperatives Produktengineering. Ein neues Selbstverstandnis des ingenieurmäßigen Wirkens.*, Heinz Nixdorf Institut, Universität Paderborn, Paderborn

Gausemeier, J. and Lückel, J. (2000): *Entwicklungsumgebungen Mechatronik - Methoden und Werkzeuge zur Entwicklung mechatronischer Systeme.* Heinz Nixdorf Institut, Universität Paderborn, Paderborn

Goldberg, D. (1991): What Every Computer Scientist Should Know About Floating-Point Arithmetic. *Computing Surveys*, March

Goldreich, O. (2000): Computational Complexity. http://wisdom.weizmann.ac.il, Last visited: 2005, Weizmann Institute of Science

Goltz, M. (2000): Product Data Controlled Engineering Workflow in the Supply Chain. In: Proceedings of ProSTEP Science Days, 13–14 September, Stuttgart

Grünwald, P. D. and Vitányi, P. M. B. (2003): Kolmogorov Complexity and Information Theory. *Journal of Logic, Language and Information*, 12, pp. 497–529

Günterberg, B. and Kayser, G. (2004): SMEs in Germany: Facts and Figures (Report). Institut für Mittelstandsforschung Bonn, Bonn

Håstad, J. (1999): Complexity Theory (Unpublished Work). Royal Institute of Technology, Stockholm

Howe, D. (2006): The Free On-line Dictionary of Computing (FOLDOC). http://www.foldoc.org/, Last visited: 2006

IEEE (1991): *IEEE Standard Computer Dictionary: A Compilation of IEEE Standard Computer Glossaries.* Institute of Electrical and Electronics Engineers, New York

Janocha, A. T. and Gandyra, M. (1997): Richtungsweisender Einsatz der Komponententechnologie für CAx-Systeme. In: Proceedings of VDI-Tagung "Neue Generation von CAD/CAM-Systemen: erfüllte und enttäuschte Erwartungen", 28–29 October, München

Jeckle, M. (1999): Modellaustausch mit dem OMG XML Metadata Interchange Format (XMI). In: Proceedings of International Knowlege Technology Forum '99, 16–18 September, Potsdam

Jesko, D. and Endig, M. (2000): Integration von Prozeßmodellierungsmethoden im Rahmen einer Prozeßzentrierten Entwurfsumgebung. In: Proceedings of Komponentenorientierte betriebliche Anwendungssysteme, Wien

Kilb, T. and Arnold, F. (1998): Data Management in Distributed CAx Systems. In: Proceedings of ProSTEP Science Days '98: Product Data Technology - Facing the Future, Wuppertal

Kotonya, G. and Sommerville, I. (1998): *Requirements Engineering: Processes and Techniques.* John Wiley & Sons, Chichester

Krause, F.-L., Kramer, S. and Rieger, E. (1991): PDGL: A Language for Efficient Feature-Based Product Gestaltung. *Annals of the CIRP*, Vol. 40, 1, pp. 135–138

Krause, F.-L., Tang, T. and Ahle, U. (2002): *integrierte Virtuelle Produktentstehung Abschlussbericht.* Fraunhofer IRB, Stuttgart

Lee, J. Y. (2001): Shape Representation and Interoperability for Virtual Prototyping in a Distributed Design Environment (Generic). Springer, London

Lutters, E. (2001): *Manufacturing integration based on information management.* University of Twente, Enschede

Nyhuis, P. and Wiendahl, H.-P. (2004): 3-Sigma PPC – A Holistic Approach for Managing the Logistic Performance of Production Systems. In: Alting, L., Bramley, A., Brinksmeier, E., Corbett, J., Dini, G., Kruth, J. P., Lucca, D. A., Monostori, L., Tichkiewitch, S., Ueda, K., Houten, F. v. and Weinmann, K. (Eds) *CIRP ANNALS 2004*, pp. 371–376, Technische Rundschau, Paris

OMG (1998): Object Management Group. The Common Object Request Broker: Architecture and Specification (Generic).

Pahl, G. and Beitz, W. (1993): *Konstruktionslehre: Methoden und Anwendung*. 3. edn, Springer, Berlin Heidelberg

Poole, J. D. (2001): Model-Driven Architecture: Vision, Standards And Emerging Technologies. Last visited: 2004, Hyperion Solutions Corporation

Poppendieck (2004): Component-Based Software Development. http://www.poppendieck.com/components.htm, Last visited: 2006

Rosenhead, J. (1998): Complexity Theory and Management Practice. http://human-nature.com/science-as-culture/rosenhead.html, Last visited: 2005

Sachers, M. (2001): White Paper for PDM-Integration of OEM and Supplier in the Automotive Industry. http://www.pdtnet.org/file/10918.wp_v_1_4.pdf, Last visited: 2005

Sametinger, J. (1997): *Software Engineering with Reusable Components*. Springer, Berlin-Heidelberg

Shannon, C. E. (1948): The mathematical theory of communication. *Bell System Technical Journal*, 27, pp. 379–423, 623–656

Siegel, J. (2000): *CORBA 3 Fundamentals and Programming*. 2nd edn, Wiley, New York

Sinclair, J., Hanks, P., Fox, G., Moon, R. and Stock, P. (Eds) (1987): *Collins COBUILD English language dictionary*. 1987 edn, William Collins Sons & Co Ltd, Glasgow

Soley, R. M. (2002): Model Driven Architecture: An Introduction. http://www.omg.org/mda/mda_files/Soley-MDA/MDA-Seminar-Soley.htm, Last visited: 2006, Object Management Group

Soley, R. M. (2005): Model Driven Architecture: Next Steps (Keynote lecture). In: Chen, C.-S., Filipe, J., Seruca, I. and Cordeiro, J. (Eds) *Proceedings of International Conference on Enterprise Information Systems*, Miami

Sowa, J. F. (2000): *Knowledge Representation: Logical, Philosophical and Computational Foundations*. Brooks/Cole, Pacific Grove

Sowa, J. F. (2001): Mathematical Background. http://www.jfsowa.com/logic/math.htm, Last visited: 2005

Sowa, J. F. (2004): The Law of Standards. http://www.jfsowa.com/computer/standard.htm, Last visited: 2006

Spur, G. and Krause, F.-L. (1997): *Das virtuelle Produkt: Management der CAD-Technik*. Hanser, München; Wien

Stachowiak, H. (1973): *Allgemeine Modelltheorie*. Springer, Wien

Stal, M. (1997): Componentware – von der Komponente zur Applikation. *OBJEKTspektrum*, 3, pp. 86–89

Starzyk, D. (2002): Review of Boeing Business Strategies and Requirements for OMG/STEP Harmonization Workshop. In: Proceedings of OMG/STEP Harmonization Workshop

Starzyk, D., Price, D. and Johnson, L. (1999): STEP and OMG Product Data Management Specifications: A Guide for Decision Makers. http://www.omg.org/cgi-bin/doc?mfg/1999-10-04, Last visited: 2005

Suh, N. P. (2001): *Axiomatic Design: Advances and Applications*. Oxford University Press, New York - Oxford

Turing, A. M. (1950): Computing machinery and intelligence. *Mind*, 59, pp. 433–460

VDI-Richtlinien (1993): VDI-Richtlinie 2221: Methodik zum Entwickeln und Konstruieren technischer Systeme und Produkte (Generic). Mai 1993 edn, VDI-Gesellschaft Entwicklung Konstruktion Vertrieb, Düsseldorf

VDI-Richtlinien (2000): VDI-Richtlinie 3633: Simulation of systems in materials handling, logistics and production - Integration of simulation into operational processes (Generic). VDI-Gesellschaft Fördertechnik Materialfluss Logistik

Warnecke, H.-J. (1996): *Die Fraktale Fabrik - Revolution der Unternehmenskultur.* 2. edn, Springer, Berlin-Heidelberg

Weber, C. (2005a): CPM/PDD – An Extended Theoretical Approach to Modelling Products and Product Development Processes. In: Proceedings of 2nd German-Israeli Symposium on Advances in Methods and Systems for Development of Products and Processes, 7–8 July, Fraunhofer IRB, TU Berlin/Fraunhofer-Institut für Produktionsanlagen und Konstruktionstechnik (IPK)

Weber, C. (2005b): CPM/PDD – An Extended Theoretical Approach to Modelling Products and Product Development Processes. In: Proceedings of 3rd International PhD Conference on Mechanical Engineering — PhD2005, 7–8 July, University of West Bohemia, Faculty of Mechanical Engineering, Department of Machine Design, Pilsen

Weber, C. (2005c): What is "Complexity"? In: Proceedings of International Conference on Engineering Design, Melbourne

Weber, C., Werner, H. and Deubel, T. (2003): A different view on Product Data Management/Product Life-Cycle Management and its future potentials. *Journal of Engineering Design,* Vol. 14, 4, pp. 447–464

Westkämper, E. (2000): Life Cycle Management and Assessment: Approaches and Visions Towards Sustainable Manufacturing. *Annals of the CIRP,* Vol. 49/2, pp. 501–522

Wiendahl, H.-P. (2002): Wandlungsfähigkeit - Schlüsselbegriff der zukunftsfähigen Fabrik. *wt werkstattstechnik online*, 04, pp. 122–129

Wiendahl, H.-P. and Heger, C. L. (2003): Justifying changeability: a methodical approach to achieving cost effectiveness. In: Proceedings of CIRP 2nd International Conference on Reconfigurable Manufacturing, August 2003, Michigan

Wiendahl, H.-P. and Heger, C. L. (2004): 37th CIRP international seminar on manufacturing systems: digital enterprises, production networks. In: Proceedings of CIRP International Seminar on Manufacturing Systems, pp. 1–9, Computer and Automation Research Institute, Hungarian Academy of Sciences, Budapest

Wiki (2006): Wikipedia. http://en.wikipedia.org, Last visited: 2006

Womack, J. P. and Jones, D. T. (1994): *Die zweite Revolution in der Autoindustrie Konsequenzen aus der weltweiten Studie aus dem Massachusetts Institute of Technology.* 8., durchges. Aufl. edn, Campus-Verl., Frankfurt/Main

Womack, J. P. and Jones, D. T. (2003): *Lean thinking banish waste and create wealth in your corporation ; revised and updated James P. Womack.* 1st Free Press , rev. and updated edn, Free Press, New York

Womack, J. P. and Jones, D. T. (2005): *Lean solutions: how companies and customers can create value and wealth together.* Free Press, New York

Woolfson, M. M. and Pert, G. J. (1999): *An Introduction to Computer Simulation.* Oxford University Press, Oxford

Index

Printing: Krips bv, Meppel
Binding: Stürtz, Würzburg